90 DAYS TO NORMANDY

BY KEN DAVIES

LYMINGTON ADVANCED LANDING GROUND.

Surviving blister hangar now used to store agricultural machinery and perfectly visible from the road between East End village and the Sealink ferry terminal at Lymington.
Photo credit: S. Rich.

Printed by
HASTINGS PRINTING COMPANY LIMITED
Drury Lane, Hastings, East Sussex TN34 1XW
Telephone 0424 720477

FOREWORD

Ninety Days to Normandy

The evacuation from Dunkirk, and the Battle of Britain left us as a nation feeling bruised and not a little bewildered.

The bravery of our forces was not in question, but it took that special brand of toughness of which we are proud, to consider ways and means of hitting back.

The RAF was worried about the fact that their main, largely pre-war Air Stations might get knocked out by enemy action, and set out to commission a number of emergency air strips. Even in those black days, it was realised that any attempt to strike back across the Channel would need airfields as near the south coast as possible.

With this in mind, some unlikely sites were earmarked as advanced landing grounds from which a counter attack could be launched. As the coast was know our front line, the plans to use these fields for attack were to a large extent hidden by the more obvious need for defence at that time.

Many farmers were infuriated that their land was suddenly requisitioned, because at the same time they were being pressed by the Ministry of Agriculture to produce more and more food. However, their protestations were overruled and from Kent to Hampshire out of the hundreds of sites surveyed by 1943, twenty-eight were established as small, crude and uncomfortable bases from which operations could be mounted.

In this book, we bring together under one cover a glimpse of that amazing partnership between units of the British, American and Commonwealth Air Forces.

Without their efforts from these advanced landing grounds, the invasion of Normandy in 1944 would not have been the success it was. The majority of them were used for as little as twelve weeks before being abandoned, allowing the land to be returned to farming.

Whilst no claim is made that this book is a definitive history of those momentous 90 days, it will be nostalgic for some, a basis for further investigation, and enable anyone interested to locate the sites that are a little known part of our past.

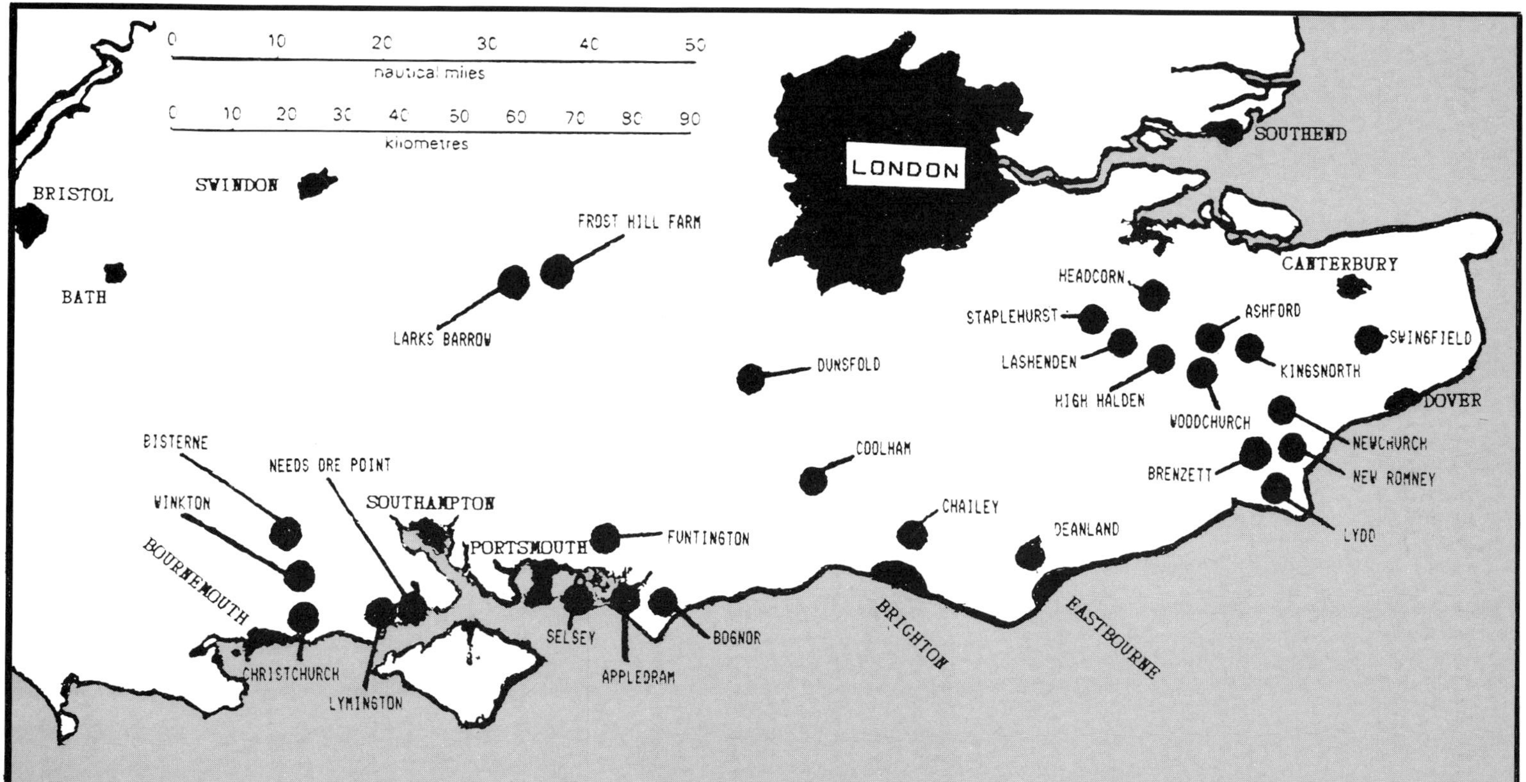

THE ADVANCED LANDING GROUNDS OF 1944.

INDEX TO ADVANCED LANDING GROUNDS

AIRFIELD	COMPUTER FILE	OS MAP REF	
ASHFORD	002	TQ 972402	
APPLEDRAM	002	SU 839018	
BISTERNE	002	SU 155030	
BOGNOR	002	SU 915005	
BRENZETT	002	TR 015280	
CHAILEY	003	TQ 370192	
CHRISTCHURCH	003	SZ 185932	**
COOLHAM	003	TQ 125222	
DEANLAND	003	TQ 523118	
DUNSFOLD	003	SQ 027361	**
FROST HILL FARM	004	SU 517525	**
FUNTINGTON	004	SU 790085	
HEADCORN	004	TQ 880460	
HIGH HALDEN	004	TQ 890397	
KINGSNORTH	005	TR 025380	
LARKS BARROW	005	SU 455505	**
LASHENDEN	005	TQ 855425	
LYDD	005	TR 015230	
LYMINGTON	005	SZ 342960	
NEEDS ORE POINT	006	SZ 402978	
NEWCHURCH	006	TR 045315	
NEW ROMNEY	006	TR 063269	
SELSEY	006	ST 865957	
STAPLEHURST	006	TQ 810432	
SWINGFIELD	006	TR 240448	
WINKTON	006	SZ 165975	
WOODCHURCH	006	TQ 945365	

**NOTE: These airfields not listed as Advanced Landing Grounds.

Britain's pride and joy, the Spitfire. This Mk9 version in pristine condition awaiting the painting on of 'invasion' stripes in 1944 before joining a wing at an advanced landing ground.

It was obvious in the last two years of the 1930s that the war to end all wars twenty years previously had failed, and the RAF were trying to shake off the lethargy that had reduced the force to a token that was good at ceremonials, but would be woefully lacking when it came to a real scrap.

Spitfires and Hurricanes were beginning to take their place in Squadrons and a number of airfields were being upgraded and equipped to support the modern aircraft. The problem was really numbers; even this was not fully appreciated when it came to the airfields, because when war broke out in 1939 it was France and the Low Countries which appeared to be forming the front line.

The full realisation of what we really lacked only dawned upon us when the German 'Blitzkrieg' tactics caused the fall of France, occupation of the Low Countries, and the forced evacuation of our army from the beaches of Dunkirk. This left the whole of the South Coast well and truly in the front line. Those established airfields, such as we had, became vulnerable overnight and the great fear was how would we defend ourselves if these were knocked out? An immediate answer was to order all the COs to locate and establish emergency airfields near their own commands.

Some weird sites were located to serve their purpose as a bolt hole if an emergency arose. Known as scatter fields these were to be ready in two weeks and it was much to the credit of the RAF personnel and others that many were, albeit without any real facilities apart from space from which to take off or land. From the aircrew's point of view these crude landing grounds gave some comfort in that they removed the fear of arriving back from an operation, short of fuel and possibly damaged, to find their own base made impossible to use due to enemy action.

As events turned out there wasn't much time to do very much more than this as the existing forces were pitchforked into the 'Battle of Britain'. The famous 'Few' with their Spitfires and Hurricanes put an end to the Luftwaffe's aspirations of a quick victory. The pilots became heroes in their lifetimes and airfields such as Hornchurch, North Weald, Biggin Hill, Kenley, Hawkinge and Tangmere were soon household names. In many cases when they returned from an operation it was only to find their home base pock-marked with bomb craters and they were glad to be able to put down safely on any strip of cleared land.

As soon as this battle was effectively won, thoughts turned to retaliation and consolidating defence. The need for more airfields was still there and none the less urgent. What had altered was that the purpose for which each one was now required had changed in emphasis. Instead of being a reserve for the 'Few', the new airfields were now to serve as stepping stones from which the planned 'Many' could launch a counter attack.

To give offence

The search began and sites for this purpose were inspected at hundreds of places south of a line from the Thames to the Severn. It was proposed that a landing was to be made on the Pas de Calais which was given the name 'Operation Hadrian' needing many temporary airfields which would only be required during the build up to and landings themselves. Because of the passing need only the bare necessities would be provided. These were stipulated as being runways, refuelling and rearming points, hardstandings plus a few blister hangars for work under cover on aircraft.

Given the general name of Advanced Landing Grounds (ALGs) these very basic airfields were to have most accommodation under canvas with a few huts for the permanent maintenance staff. By 1942 fifteen possible sites had been selected which included six favoured ones at Hadlow, Rother Levels, Marden Beeches, Withersdale, Wickham Breux and Stelling Minnis. Because of the difficulties and shortage of materials needed to get these chosen strips completed none of them were ever developed.

In the meantime Operation Hadrian had been postponed and other chosen sites from seventy-two possibles were ordered to be made ready by March 1943 for the 'big push' for a second front. By definition the establishment of an airfield was a land hungry operation which led to strong opposition from the War Agricultural Executive Committee in nearly every case. They, of course, had the important task of maximising food production which, in view of the horrific losses our merchant ships were suffering, gave the home food producers a very high priority. Such were the lengths of the discussions concerning each site that, of the seventy-two, only thirty-six were put up for acceptance. A few of these, like Dunsfold, were developed into full scale airfields, and finally a shortlist of twenty-five was drawn up of which twenty-three were actually built all along the South Coast, from Kent to Hampshire.

The Summer of 1942 saw detailed surveys carried out and a specification for ALGs was formulated. Each one was to have two runways of 3,000ft (914m) roughly at right angles to each other, laid out with Sommerfield Tracking. They were to have partial taxiways, one hardstanding for every two aircraft, two blister hangars and the minimum of other buildings.

Work started on the December of 1942 by which time a better idea of the purpose these fields would be put to and the aircraft that would be using them was taking shape. The main purpose was obviously going to be the planned invasion of Europe, the advanced landing grounds serving as a base for the Fighter Bombers with the task of softening up the coastal defences on the other side of the Channel. They would later provide front line bases for the aircraft with the task of supporting the invading forces themselves.

Eleven of those under construction were allocated to the American Air Forces and in all cases the specification upgraded to allow a main runway of 4,800ft (1,463m) and a second one of 4,200ft (1,280m). These could be shorter by 200ft, if they were aligned into the prevailing wind.

Target date for completion was still set for March 1943 so the Royal Engineers' Construction Group and the RAF were helped by the American Engineer Aviation Battalions and all soon started clearing and grading the sites, laying out the meagre facilities. Roads were closed or diverted, hundreds of trees felled, hedges and ditches removed and small hills simply levelled. It was perhaps fortunate that the build up of the Tactical Air Force was behind schedule because, in spite of very real efforts, the British weather and a shortage of materials delayed the completion of the first sites until late in the Spring of 1943.

The next five months look like chaos if you try to follow the movements of various squadrons in and out of these crude airfields. In fact it probably seemed worse for those actually having to fly in, operate without adequate facilities, and maintain a war-winning standard ready for the 'Big Push' that everybody knew was coming, but nobody really knew where or when the strike was to be made.

In fact, during those hectic few months the lessons learned were enough to enable the two dozen airfields to play a part in a successful return to Europe that would not have been possible for the allies without the fields' existence. The problems, limitations and dangers of operations from this type of unsophisticated airstrip, having been learned, were to a large extent eliminated when all the sites were closed in the Autumn of 1943 and the various units were posted back to airfields with a better weatherproofing. In any case they were never intended to operate in winter conditions, which enabled the Engineer units to move back in and remedy the serious shortcomings that had been revealed.

Probably the two worst features that had to be put right were the wear and tear on the Sommerfield Tracking that formed the runways, and the lack of accommodation and parking for the support facilities that kept the squadrons mobile.

Various other reinforcing materials were tried to replace the Sommerfield Track, the best being the American Pierced Steel Planking, but this was in short supply, expensive and difficult to transport and lay. Another American product, Bar and Rod Track, was lighter than PSP, but more difficult to manufacture. The main use found for this was on taxiways especially on bends which had to be reinforced where the wear was greatest. Impregnated hessian was tried as a possible way to waterproof the surface but in wet weather it took on all the qualities of a skating rink.

In the end, trials of a British product solved many of the problems revealed by the Summer's operations. A three inch square steel mesh produced was used, manufactured in rolls by British Reinforced Engineering Company Ltd. This proved easy to transport, easy to lay and apart from the frightening tendency to build up a steel 'bow wave' in front of aircraft taking off seemed to do the job well. The 'bow wave' effect was cured by extra tensioning and picketing when laying the runways, and for practical purposes cured the problem.

Other improvements in hardstanding, provision for support vehicles, extra blister hangars and extension of taxiways went ahead with all speed. The Americans carried out work on the fields allocated to them, installed transportable hangars, and reinforced the runways to stand up to the terrible beating given them by the P-47 Thunderbolts. It was a great accolade to the unsung forces who worked through that winter of 1943-44 when operations were able to start on April 1, 1944, as planned. Considering the conditions and the lack of experience the pilots had in this type of operation, accidents were mercifully few and the life of the ALGs short. Missions were intensive enabling a vital contribution to the invasion to be made by the 'Many' in the best traditions set by the 'Few' three years before.

To prove a point

I have picked out two Advanced Landing Grounds as typical, one operated with RAF aircraft, the other with American aircraft and personnel. These two have been given the full treatment of a start to finish story of their existence. To complete the picture, all are listed

A Typhoon 1B of 440 Squadron thought to be seen in this photograph at Funtingdon.

Spitfires beign prepared to join No. 83 support group at Bognor ALG.

with dates of opening and closing, squadrons and aircraft based on them and their geographical position given as an OS reference. Highlights of operations from them have been mentioned in a little detail where they were of importance or interest.

Like other Advanced Landing Grounds Appledram was laid out with two Sommerfield Track runways. The site chosen was on flat ground beside the Chichester Channel near Dell Quay within sight of the Cathedral, only a few miles from the famous Battle of Britain airfields of Tangmere and Westhampnett (better known nowadays as Goodwood).

After the flurry of activity, of contractors laying the mesh runways and generally clearing the site, it was opened in May 1943 when three squadrons of Typhoons (Nos 175, 181 & 182), which formed one of the new Fighter Bomber Wings moved in from Lasham to commence operations. Of the three, 181 and 182 Squadrons were experienced in fighter bomber type operations, but 175 having converted to Typhoons a month before had much to learn.

The first operations from Appledram were made against Abbeville Airfield on 12 June, 1943 and followed by intensive operations against all forms of communication, airfields and other military installations in Northern France.

It wasn't only aircrew who were learning what it was like to operate with minimum facilities, all support staff were accommodated in tents, the aircraft were dispersed around the site, and maintenance had to be carried out in the open, or at best under canvas screens. In spite of the difficulties, operations were carried out with considerable success, much was learned, and at Appledram there was only one serious accident in that time to mar what looked like being a complete success.

Advanced Landing Grounds were never conceived as being anything but a very temporary measure to support the one major operation of the Allied assault on Europe. Appledram had taught the Wing stationed there a lot and to enable these lessons to be used to the best advantage the site was reverted to standby status in July 1943. This was to allow improvements to be carried out in the winter months which were to include those lessons learned from the other ALGs. These included those of operational procedures by the newly formed Wings and so 175 Squadron was moved to Lydd on 1 July, 1943 and 181 and 182 transferred to New Romney a day later.

A fully loaded Typhoon was no lightweight, so the next eight months saw a number of improvements being made to the site which mainly consisted of improving the load bearing qualities of the Sommerfield Mesh Runways, extending taxiways, and putting down hardstanding bays. A few blister hangars were also provided, but little if anything was done about accommodation. Eventually when ground staff did return, some limited requisitioning of local farm buildings took place, not a perfect arrangement, but better than damp tents!

The real thing

Re-opening on 3 April, 1944 Appledram immediately played host to 310 Squadron with Spitfire Mk Vbs which moved in on that day, joined by 312 Squadron with Spitfire IXs and No 313 with their Spitfire Vbs a day later. The build up to D-Day had begun! These three Czech units then got down straight away to a daily round of intruder sweeps, escorting bombers and attacking 'Noball' (flying bomb) sites making the small airfield a hive of activity for the next two and a half months.

On D-Day itself the three units were operating as a Wing over the Beachhead, but by standards expected enemy activity was light; they only found a few FW 190s to engage, but quickly disposed of five of these. This wing stayed at Appledram until 22 June, when all three squadrons moved the few miles east to Tangmere to provide extra defence against the increasing Flying Bomb attacks.

A few days later on 28th June, 302, 308 and 317 Polish Squadrons were based here beside Chichester Harbour with Spitfire IXs. This Wing was soon on the offensive, attacking German positions as and when called upon by the Army. This represented a

A line up of Typhoons waiting a scramble on a unidentified airfield in 1944. It is of interest to note the liaison DH Domonie biplane visible beneath the wing of the leading Typoon.

A Mustang III of 316 (Polish) squadron. Whole wings of these aircraft flew from ALGs in 1944. Unfortunately this particular one crashed on take-off from Andrews Field in March 1945.

technique honed to perfection by the Allied air forces co-operating with ground forces. This type of operation was carried on until the final victory. The advance of the Allied forces on the Continent meant that Advanced Landing Grounds rapidly became backwaters out of range of anywhere that real actions were taking place. The result of this was that after a stay of only two and a half weeks the Polish Wing was moved to Ford on 16 July in preparation for being moved over the Channel to where the action was.

Beginning of the end

Very little happened after that at Appledram, it had served its purpose, being duly derequisitioned in November 1944. Two Flights were sent from 5027 Works Squadron in January 1945 with the task of dismantling such fixtures and fittings that there were. They removed the blister hangars and lifted the Sommerfield Tracking so that the land could be returned to agriculture. In this way this patch of land beside the Chichester Channel was able to continue providing a vital contribution to the war effort with food production.

Of all the 23 Advanced Landing Grounds Appledram did not see the most action, nor was it the busiest. On the other hand it was typical of them all, its contribution of 'Ninety Days to Normandy' as an Air Base, being an essential part of the opening of the 'Second Front' by the Allies. The chances are that forty-five years later Appledram, like so many other Advanced Landing Grounds would have been forgotten were it not for the establishment of a small, but very interesting museum on the site, with hundreds of exhibits showing clearly what a D-Day airfield was like. Replica and model aircraft with many actual relics of those traumatic days in 1944, all go to recreate a very real atmosphere from the past.

Thunderbolts from Lymington

The Advanced Landing Ground known locally as Wallhampton, but in official records as Lymington, was in common with others built with two wire mesh runways, one east to west 1,600 yards long and the other of 1,400 yards running north to south, to serve for D-Day operations.

The site chosen was a mile to the east of Lymington, not far from the present Isle of Wight ferry terminal. Three or four blister hangars were built to provide some cover in the rather exposed position overlooking the Solent only forty-five feet above sea level.

As far as a study of records tells us the first aircraft to use Lymington were three squadrons of the USAAF who flew in during March 1944. These aircraft were P-47 Republic Thunderbolt fighter bombers of the 50th Fighter Group (10, 81 and 313 Squadrons) under the control of the IXth Tactical Air Command.

After a few weeks of intensive training their first operation from Lymington was flown on 1 May, 1944 after which non-stop missions were carried out in daylight hours by all three squadrons attacking enemy communications on the Continent.

These attacks were of increasing frequency over the six weeks prior to D-Day and continued until landing strips became available on the continent as the Allies advanced. When this situation came about the whole group flew from Lymington to landing grounds in Northern France where they could give a closer support. The airfield at Lymington was never used again after the Thunderbolts left for France. It had served its purpose and early in 1945, after the mesh runways and taxiways had been removed, the site returned to its former agricultural use.

Unlike Appledram it would take a very close inspection today to tell that at one time this site was for ninety days a scene of an intense aerial activity that played a vital part in our invasion of Europe. The keen observer will note that at one point where the north-south runway used to cross a minor back road from Beaulieu to Lymington the hedge has been replaced by posts and wire. Just to the north of this road beside another unclassified road is one of the old blister hangars in remarkably good condition and now being used to store farm equipment.

US Air Force Republic P47Ds (Thunderbolt) fighter bombers typical of hundreds that saw service from ALGs.

A Tempest V of 3 Squadron being re-armed at 'Newchurch' ALG prior to a sortie in summer 1944.

Not all the ALG sites were identical. Obviously the geographical position gave each one different characteristics; some were much hated and downright dangerous for any form of living. Their temporary nature and the fact that they were only conceived as staging posts with an operational life of a few months essential to the prosecution of the war were the justification for their use.

The nature of the tactics dictated that they would all have to be within a few miles of the South Coast to enable maximum operational time over the Channel and it will be seen from the map that the 23 ALGs were in four main groups, around Ashford, New Romney, Chichester and the New Forest areas. There were the few oddballs and a number were intended to be Advanced Landing Grounds, but for one reason or another were never used for that purpose. Little or no information is readily available about some of these, while others like Dunsfold blossomed into fully fledged airfields and are still in use today.

There were also a few sites that were already in use for other purposes, like Christchurch (known locally as Sommerfield), which had been in use pre-war as a private airfield and by 1943 was accommodating Airspeed's manufacturing facility moved from Portsmouth. Suddenly it had Sommerfield tracking laid down and was pressed into service as an Advanced Landing Ground to be used by USAAF P-47 Thunderbolts over the invasion period.

Although some of these ALGs were opened in 1943 for training purposes, the real action took place from April to June 1944, the three months immediately prior to D-Day itself.

The following list gives the location and dates of opening and closing, where known, of each field. As far as possible the dates listed as opening are the ones on which the first flying took place and closing dates given are the ones on which the sites were derequisitioned. Where this is in any doubt the date of the last known unit leaving there is given and further unusual use or development is told in the text that follows.

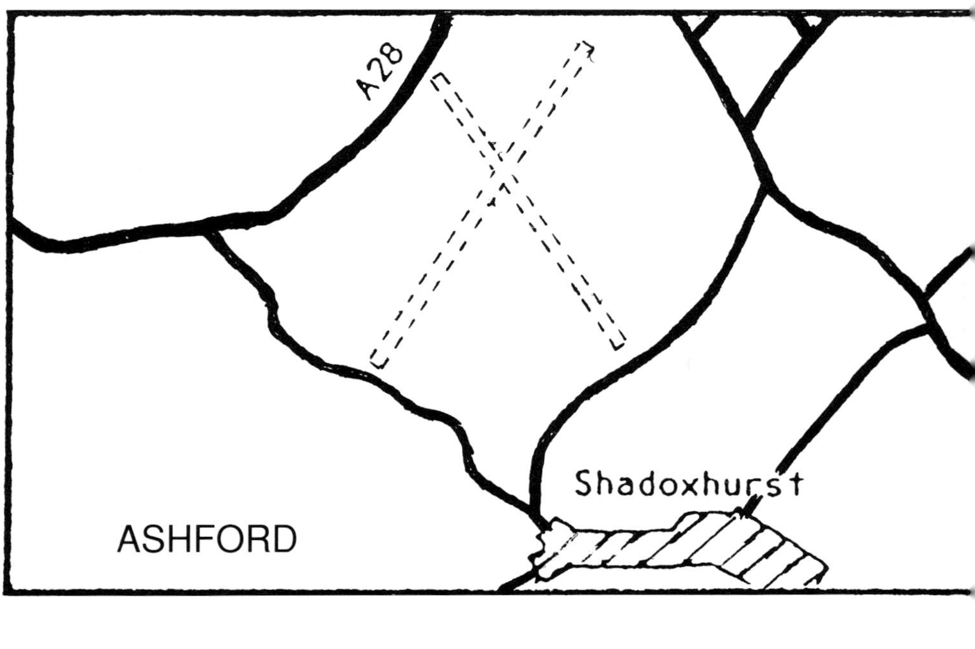

Ashford (Great Chart), Kent

Unit	Arrived from	Date	Left for	Date	Type
414 Sqn RCAF	?	13/08/43	???	??/10/43	Mustang I
430 Sqn RCAF	?	13/08/43	???	?0/10/43	Mustang I
?? Wing RAF	Kingsnorth	??/10/43	???	??/10/43	Spitfire Vs
512 Sqn USAF	?	04/04/44	France	27/07/44	P-47 Thunderbolt
513 Sqn USAF	?	04/04/44	France	27/07/44	P-47 Thunderbolt
515 Sqn USAF	?	04/04/44	France	27/07/44	P-47 Thunderbolt

Airfield	OS Map Ref	Opened	Closed	Remarks
Swingfield	TR240448	15/06/43	28/04/45	6m N of Folkestone
Woodchurch	TQ945365	28/07/43	18/09/44	5½m SW of Ashford
Newchurch	TR045315	02/07/43	13/12/45	7m SE of Ashford
New Romney (Honeychild)	TR063269	02/07/43	01/11/44	1½m N of New Romney
Lydd	TR0 5230	25/06/43	13/12/44	2½m NW of Lydd
Brenzett (Ivychurch)	TR015280	??/09/43	??/12/44	3m NW of New Romney
Ashford (Great Chart)	TQ972402	11/08/43	15/09/44	2½m SW of Ashford
Headcorn	TQ880460	20/08/43	09/07/44	8m W of Ashford
Lashenden	TQ855425	06/08/43	†29/06/45	9m SE of Maidstone
Kingsnorth	TR025380	01/07/43	20/07/44	3m SE of Ashford
High Halden	TQ890397	13/04/44	15/09/44	7½m SW of Ashford
Staplehurst	TQ810432	06/08/43	18/01/45	8m SE of Maidstone
Deansland	TQ523118	§01/04/44	12/01/45	4½m NW of Hailsham
Chailey	TQ370192	01/04/43	20/01/45	3m E of Burgess Hill
Coolham	TQ125222	01/04/44	01/08/44	6m SW of Horsham
Bognor	SU915005	01/06/43	26/09/44	1½m NW of Bognor
Selsey	ST865957	30/05/43	19/08/44	5m S of Chichester
Appledram	SU839018	02/06/43	06/11/44	2½m SW of Chichester
Funtington	SU790085	15/09/43	13/12/44	5m W of Chichester
Needs Ore Point	SZ402978	10/04/44	??/11/44	3½m S of Beaulieu
Lymington	SZ342960	??/03/44	25/06/44	1m E of Lymington
Wington	SZ165975	04/04/44	06/07/44	1½m W of Sopley
Bisterne	SU155030	03/03/44	23/06/44	1½m S of Ringwood
Dunsfold **	SQ027361	†10/02/42	†IN USE	6m SE of Godalming
Frost Hill Farm **	SU517525	§ ??/06/40	§ ??/12/45	1½m N of Overton
Larks Barrow **	SU455505	§ 10/12/42	§ ??/01/44	1½m NW of Whitchurch
Christchurch**	SX185932	§ 11/04/44	§ 30/06/44	1½m NE of Christchurch

NOTE: ** These sites are special cases, see text.
† See text for later development of site.
§ See text for clarification of dates.
All the above ALGs have been listed roughly in an East to West order to enable easy reference for those wishing to explore further. The text that follows will give a brief history of each field and list squadrons and aircraft types that operated from them.

The locals knew this airfield as Great Chart. An early start was made on the constructio on the site in the apex of A28 and a minor road to Stubb's Cross. When 'Operatio Hadrian' was cancelled, anti-aircraft guns were installed and two Mustang squadrons of th RCAF flew in with the widest possible brief to fly reconnaissance missions over th Channel. When the Canadians left the RAF took over with a Spitfire Wing whose main jo was to fly a cover for bombers and carry out diversionary sweeps, but they only stayed fo ten days and moved out when the field was handed over to the United States 9th Air Forc for fighter bomber operations.

They improved the field by building and completing the taxiways, adding to th hardstandings and extending the main runway. The group became operational early in th May of 1944 and set about the softening up of enemy defences and communications i Northern France. On D-Day itself they were kept very busy over the beaches and sadly los five aircraft during these operations after which they were reverted to reconnaissance flight and further attacks on enemy ground forces.

513 Squadron had the distinction of being able to claim the first destruction of a V- flying bomb by the USAAF. They also were amongst the first units to arm with 127mn rockets which were to prove such a success against tanks and armoured vehicles.

Bad weather, and late completion of Advanced Landing Grounds in France delayed th Wing's planned move across the Channel late in June by about a month and Great Char was released to No 42 group for use as a storage area in September 1944, a task i performed until returned to agriculture in early 1945.

Bisterne ALG was completed in September 1943 on land to the east of the River Avor south of Ringwood. The minimum of facilities were provided, but by standards of som other Advanced Landing Grounds they were lavish. Taxiways and hardstandings and PSI marshalling areas were laid down and some temporary hangars built.

Bisterne (Hants)

Unit	Arrived from	Date	Left for	Date	Type
371 FG, USAAF	USA	??/03/44	France	??/07/44	P-47 Thunderbolt

Bisterne Advanced Landing Ground

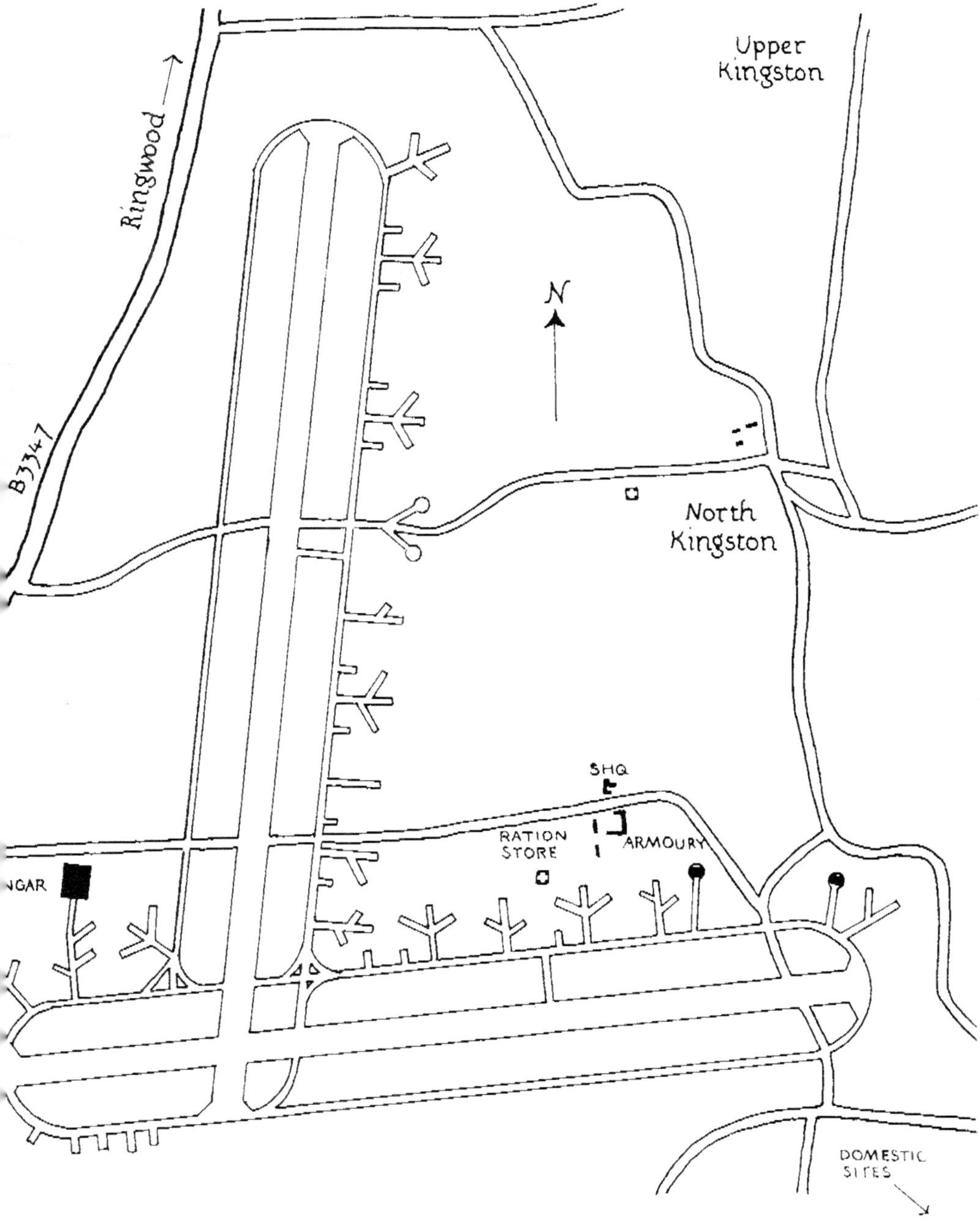

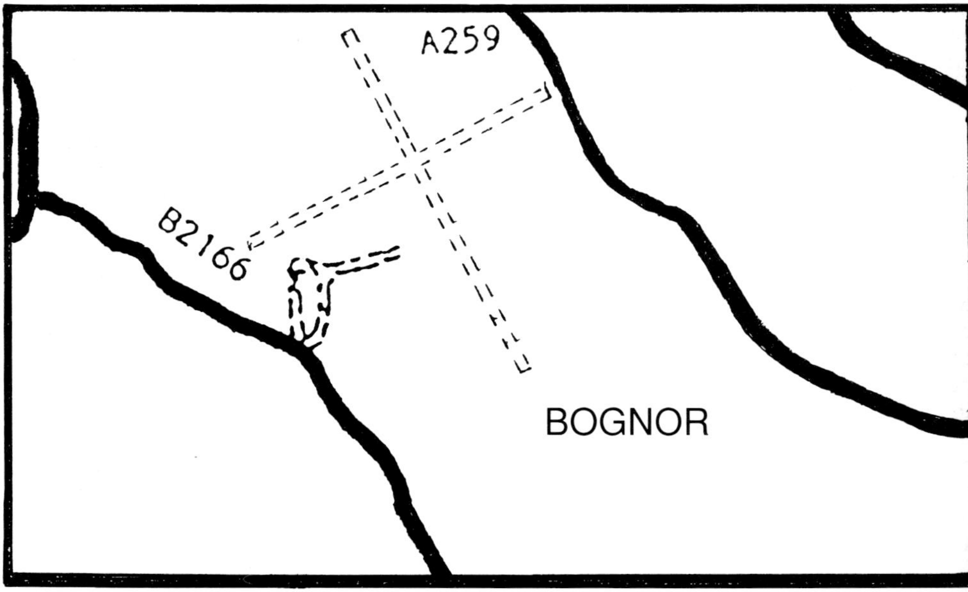

Bognor (West Sussex)

Unit	Arrived from	Date	Left for	Date	Type
122 Sqn RAF	Eastchurch	01/06/43	Kingsnorth	01/07/43	Spitfire Vc
602 Sqn RAF	Fairlop	01/06/43	Kingsnorth	01/07/43	Spitfire Vc
19 Sqn RAF	Gravesend	02/06/43	Newchurch	02/07/43	Spitfire Vc
66 Sqn RAF	North Weald	31/03/44	Southend	22/04/44	Spitfire IX
331 Sqn RAF	North Weald	31/03/44	Tangmere	22/04/44	Spitfire IX
332 Sqn RAF	North Weald	31/03/44	Tangmere	21/06/44	Spitfire IX
83 Group support unit					
Spitfires, Mustangs and Typhoons from Redhill					
Ambulance Ansons of 1310 Flt					

The surprising thing was that apart from the odd visiting aircraft the airfield was not used until April 1944 when the 371st Fighter Group of the IX Tactical Air Force moved in. Their Thunderbolts ruined the Sommerfield tracking and they had to be moved to nearby Ibsley, while hasty repairs were made to enable the planned harrying missions to be flown over France.

There were casualties, but missions increased in intensity as D-Day grew nearer. On the day itself the Thunderbolts flew dive bombing missions in support of the troops landing on the beachheads as well as the paratroops behind the enemy lines and continued with these operations until they were able to move across the Channel in July 1944. Once the Americans left, Bisterne was left to rot away. The mesh runways were lifted in the early part of 1945, the site being returned to agriculture leaving very few traces to show where this ALG was.

A municipal airport had been proposed for Bognor as early as 1928, plans for which never came about. The land was eventually covered by a housing estate and it is quite amusing to find that in 1942 when surveyors were looking for a site on which to build an ALG they found that the proposed airport site had been built over. Nobody seemed to know and certainly nobody told them!

In the end they had to choose another site which unfortunately was on good farm land to the North-West of the town. The Ministry of Agriculture violently opposed this choice, but in the end lost out to the Air Ministry and the airfield was laid out with the usual Sommerfield Mesh runways. The airfield was opened in June, 1943 under the general command of Tangmere for use as an exercise strip.

During the winter of 1943-44 improvements were made in accordance with the lessons learned from other similar sites and an operational wing moved in during March, 1944. Squadrons 331 and 332 were manned by Norwegians and achieved considerable success destroying six German aircraft on one particular operation against an airfield near Paris in the April of 1944. On D-Day and the weeks that followed the Wing saw all the action it could handle, beach head patrols, bomber escort, and attacks on V-l sites were all included in the actions they undertook from Bognor prior to moving over to Tangmere.

The increasing frequency of V-l attacks on London forced the move on large numbers of aircraft normally based at Redhill. Bognor had to host large numbers of Spitfires, Typhoons, Mustangs and an Anson of an Air Ambulance Flight. This Anson was mainly allocated the task of ferrying blood plasma to France, a job eventually taken over by a Dakota.

With all this activity from the crude facilities of an ALG, accidents did happen and one Mustang crash landed on August 8, 1944, while another was making a forced landing because of an engine fire when it collided with another aircraft waiting to take off, killing the pilot. All the units were moved to Thorney Island in September, 1945 and Bognor ALG was stripped of all fixtures and fittings before being returned to its owners in the early part of 1945.

Appledram (West Sussex)

Unit	Arrived from	Date	Left for	Date	Type
175 Sqn RAF	Lasham	02/06/43	Lydd	02/07/43	Typhoon Ib
181 Sqn RAF	Lasham	02/06/43	New Romney	02/07/43	Typhoon Ib
182 Sqn RAF	Lasham	02/06/43	New Romney	02/07/43	Typhoon Ib
310 Sqn RAF	Southend	03/04/44	Tangmere	22/06/44	Spitfire Vb
312 Sqn RAF	Mendeslham	04/04/44	Tangmere	22/06/44	Spitfire Vb
302 Sqn RAF	Chailey	28/06/44	Ford	16/07/44	Spitfire IX
308 Sqn RAF	Chailey 28/06/44		Ford	16/07/44	Spitfire IX
317 Sqn RAF	Chailey 28/06/44		Ford	16/07/44	Spitfire IX

The full story of operations from this airfield has been told already as it is typical of the majority of other ALGs from which the RAF Squadrons operated.

Although this airfield was completed in 1943 it was let out for grazing during the winter of 1943-44. As with most of the other ALG sites there was opposition from the Ministry of Agriculture, but in Chailey's case the added blow was the demolition of the 'Plough Inn' which brought about unasked for support for the Agricultural Ministry from the local people who thought that the destruction of their 'local' was too much to ask even for the war effort.

A wing of Polish manned Spitfire IXs used the field for May and June 1944 flying operations aimed at softening up enemy communications prior to D-Day, low level on the day itself and anti-VI flights afterwards. Much of their operations were plagued by wear and tear on the Spitfire's tyres from the uneven Sommerfield tracking. In spite of this, accidents were rare and the Wing moved on to Appledram when the general move forward was possible.

When they left Chailey had served its purpose and apart from a Lysander from 161 Squadron that crash landed there in October 1944 no other flights were made from this field. The site was derequisitioned in January 1945 and cleared, leaving only a few gaps in the hedgerows as evidence that it was for two months a hive of aviation activity.

There has, in fact, been a whole book written and published about Christchurch airfield which was opened as a flying club in 1935. When the Second World War broke out in 1939 it was closed down only to be re-opened six months or so later as a suitable site for the Air Ministry Research Establishment from Swanage and its Special Duties Flight. In 1940 a Shadow Factory was built along the field's western edge and this was soon occupied by Airspeed Ltd for the production of Oxford training aircraft.

Conversion of Spitfires into Seafires and assembly of Horsa gliders added to activities which attracted unwelcome attention from German bombers on several occasions. A Sommerfield tracking runway was added in 1943 and a Wing of Thunderbolts of the 405th FG 84th Fighter Wing of the IX Tactical Air Command USAAF moved in during April 1944.

These heavy fighter bombers were accident prone due to the very uneven surface of the runway at this airfield. One actually crashed into some adjacent bungalows and exploded, killing 19 and causing considerable damage around the area. Three crashed on one day and it was common for the bombs to come off during take off because the aircraft bounced badly as they gathered speed over the bumpy runway.

In spite of this, the Americans mounted many operations from Christchurch in the general softening up of enemy communications prior to and after the invasion of France on D-Day. The Thunderbolts moved across the Channel on June 30, 1944, and Christchurch reverted into an extremely active manufacturing site. Airspeeds and De Havillands

Chailey (East Sussex)

Unit	Arrived from	Date	Left for	Date	Type
302 Sqn RAF	Deanland	26/04/44	Appledram	28/06/44	Spitfire IX
308 Sqn RAF	Deanland	26/04/44	Appledram	28/06/44	Spitfire IX
317 Sqn RAF	Deanland	26/04/44	Appledram	28/06/44	Spitfire IX
Lysander from 161 Sqn force known to have crash landed on 18/10/44					

Christchurch (Dorset)

Unit	Arrived from	Date	Left for	Date	Type
405 FG, USAF		??/04/44	France	30/06/44	P-47 Thunderbolt

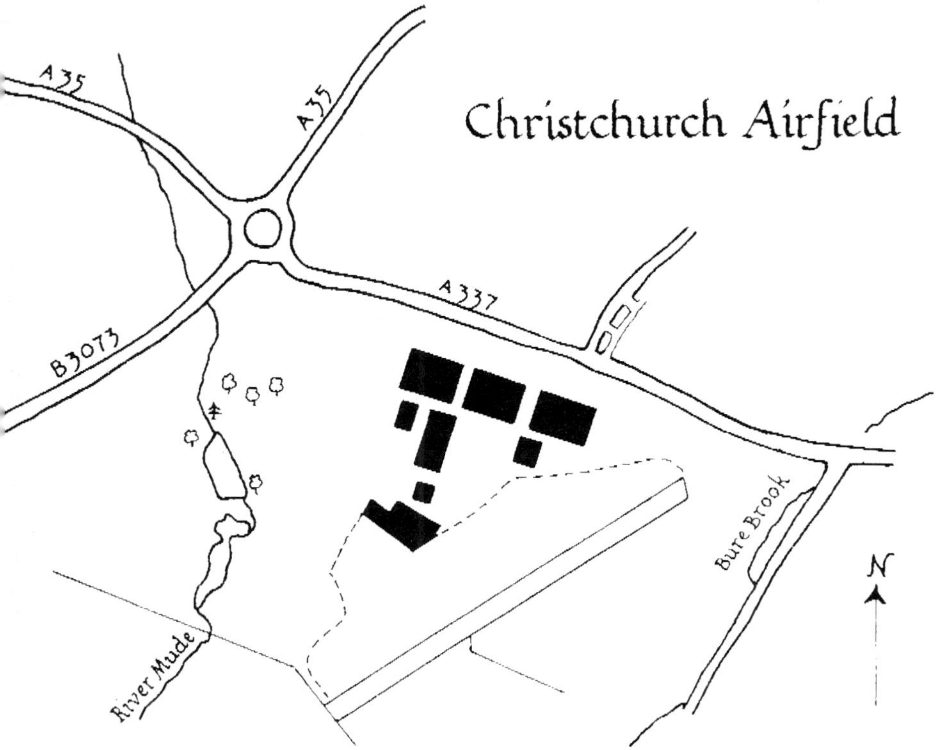

continued with this well after the war was over and the airfield, having served for a few months as an Advanced Landing Ground, was finally closed down in 1962. Today, only a few roads with aviation names and memories are left of this very active site that contributed much to the aircraft industry in peace and war.

The expected opposition from the Ministry of Agriculture was made as soon as the ALG was proposed at Coolham. The opposition must have been very strong here because the unusual step was taken of re-aligning the perimeter track to save a farmhouse in 1943 although this didn't stop the demolition of other buildings. After all this preparation the site was cleared for grazing and not brought into use until the arrival of a Mustang Wing on April 1, 1944.

The usual build up to the invasion was supported by operations deep into France by dive bombers from Coolham. By virtue of accommodation in nearby buildings it was felt to be one of the more comfortable ALGs on which to be based.

One remarkable rescue story about a Polish Warrant Officer is worth telling as it is probably only one of the many that came to pass during the hectic months during Operation Overlord. WO Tamowicz was shot down and injured when crashing on marshland near a partially constructed airstrip south of Cherbourg. He was seen by

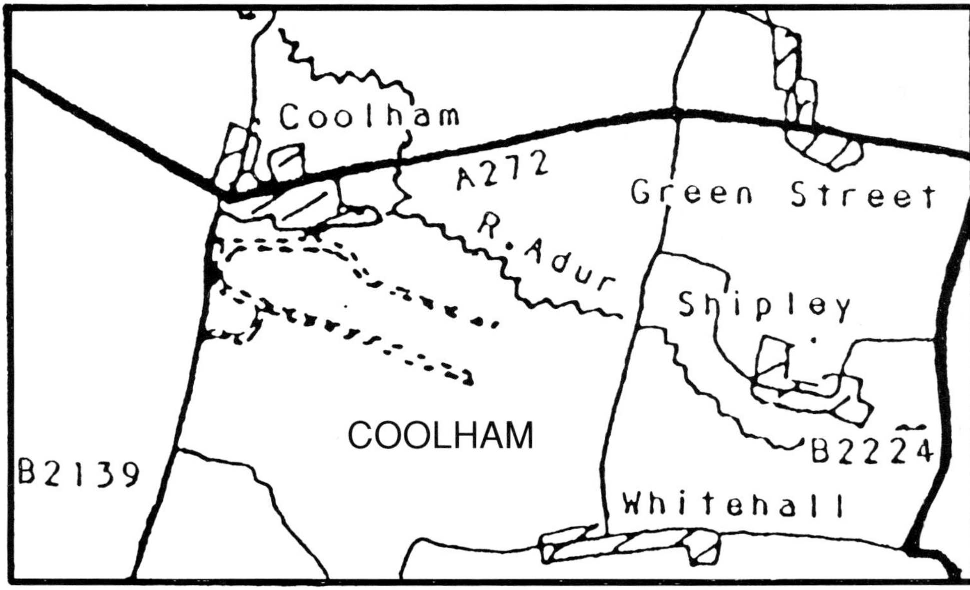

Coolham (West Sussex)

Unit	Arrived from	Date	Left for	Date	Type
306 Sqn RAF	Heston	01/04/44	Holmsley Sth	22/06/44	Mustang III
315 Sqn RAF	Heston	01/04/44	Holmsley Sth	26/06/44	Mustang III
129 Sqn RAF	Llanbeder	06/04/44	Holmsley Sth	22/06/44	Mustang III
222 Sqn RAF	Selsey	30/06/44	Funtington	04/07/44	Spitfire IX
349 Sqn RAF	Selsey	30/06/44	Funtington	04/07/44	Spitfire IX
485 Sqn RAF	Selsey	30/06/44	Funtington	04/07/44	Spitfire IX
USAF Liberator landed on 19/01/45 for repairs; flown out 26/01/45					

Deanland (East Sussex)

Unit	Arrived from	Date	Left for	Date	Type
302 Sqn RAF	Northolt	01/04/44	Southend	12/04/44	Spitfire IX
308 Sqn RAF	Northolt	01/04/44	Chailey	26/04/44	Spitfire IX
317 Sqn RAF	Northolt	01/04/44	Chailey	26/04/44	Spitfire IX
64 Sqn RAF	Coltishall	29/04/44	Harrowbeer	23/06/44	Spitfire IX
234 Sqn RAF	Bolt Head	29/04/44	Predannack	19/06/44	Mustang III
91 Sqn RAF	West Malling	21/07/44	Biggin Hill	07/10/44	Spitfire XIV
322 Sqn RAF	West Malling	21/07/44	Fairwood Common	10/10/44	Spitfire XIV
345 Sqn RAF	Shoreham	16/08/44	Biggin Hill	18/10/44	Spitfire IX

Squadron Leader Horbaczewski who promptly landed, took a Jeep, drove to the crash site, waded out and carried the injured WO back to the Jeep and so to his Mustang. The look on the ground crew's faces must have been a picture when he landed back in England and the two of them emerged from the none too generously sized Mustang's cockpit. Coolham's life was short but hectic. Even so before it was finally cleared an American Liberator bomber made a forced landing there in January 1945. Luckily there was sufficient of the runway left for it to take off again after repairs. After the tracking was lifted the site soon returned to its more peaceful farming activities.

The pattern of events for Deanland was similar to many of the other ALGs. It was, in fact, completed in 1943 but not officially opened until April 1944. Actually almost as soon as the first runway was laid down a wounded Spitfire pilot spotted it and landed there before he passed out.

In the September of that year what must have been its largest visitor landed. This was a badly damaged B-17F, Flying Fortress, which became lost and was dangerously short of fuel on returning from a daylight raid on Stuttgart.

The first official residents were squadrons of a Polish Wing of Spitfires, who were undergoing mobility training having recently converted to fighter bombers. With typically Polish flair they threw themselves into the job of learning how to hit the enemy hard by the simple process of overstepping orders and attacking at every available opportunity. Other units moving in later provided escorts for Boston, Marauder and Mitchell medium bombers which were attacking everything that moved and quite a lot that didn't in Northern France prior to D-Day.

One of Deanland's less endearing features was that it was directly under the flight path of most of the V-1s launched against the London area which made the constant firing of the anti-aircraft defences so noisy that sleep was difficult. In spite of this the units based there over the initial stages of the invasion provided every conceivable type of support operation for the invading forces.

The final move out to France was not complete before October, 1944, and it was a month later when the site clearance started. Some of the tracking now forms roadways as part of the Deanland Wood Caravan Park and bits of the hardstandings are used for storage of farm equipment.

Frost Hill Farm

The site at Frost Hill Farm was first surveyed in 1940 and was later built as a reserve ALG. It was occasionally used between 1940 and 1941 by aircraft from Odiham, but was not really used as anything but an emergency landing ground or a scatter site as they were then known. It was resurveyed as a full ALG in July 1942 and mesh runways were installed but it had no buildings a year later. As events turned out, it was never used.

A fine flying view of an American Air Force Replublic Thundercat P47D. Typical of the hundreds that were seen over the skies of southern England in 1944.

Funtington

To list all the activities of the Wings that operated from Funtington is quite beyond the scope of the space on these pages. From the point of view of the number of units based there for varying periods between the first of April 1944 and the middle of August in that year, and the endless operations they carried out it must rate as the busiest of all the Advanced Landing Grounds.

Fighter escorts, intruder raids, and dive bombing sorties were the norm for all the various types of aircraft using the site. Victories in the form of enemy aircraft destroyed are proudly claimed by Wings of Typhoons and Spitfires operating from Funtington, which added to the morale of other units with less tangible results to show for their efforts. Just being part of a winning team worked wonders.

Among the more notable persons to fly from Funtington in 1944 was Wing Commander J. E. (Johnny) Johnson, DSO, DFC and Bar, who led 441 and 443 Squadrons on a number of bomber escort operations. Two days after their arrival, 'Johnny' led 24 of their Spitfires on an escort mission with Flying Fortresses of the USAAF and engaged a number of FW 190s, shooting down six of them. Unfortunately, this operation was not without loss, but was typical of hundreds carried out from the ALGs.

The last units to operate from Funtington were the four squadrons of 132 Wing. They were also engaged in escort work before moving out on 12 August 1944 to Ford on their way to France. This left the field suddenly quiet after four-and-a-half months of intensive flying. The silence must have nearly hurt those who had got used to the activity and almost felt a part of it, but the site was derequisitioned in the December and agriculture resumed after clearance of the land.

Headcorn

At first Headcorn was known as Coldharbour when originally surveyed by Fighter Command, but rejected because of construction difficulties. A little later it was decided to go ahead with an ALG for two main reasons, the natural camouflage possible and the strong opposition to some of the other sites being considered. When the work did go ahead it was completed by early June 1943 in spite of the difficulties encountered.

The first two squadrons to fly in were Canadian who soon became engaged in escort of bombers over Northern France. In the September of that year the airfield staff were

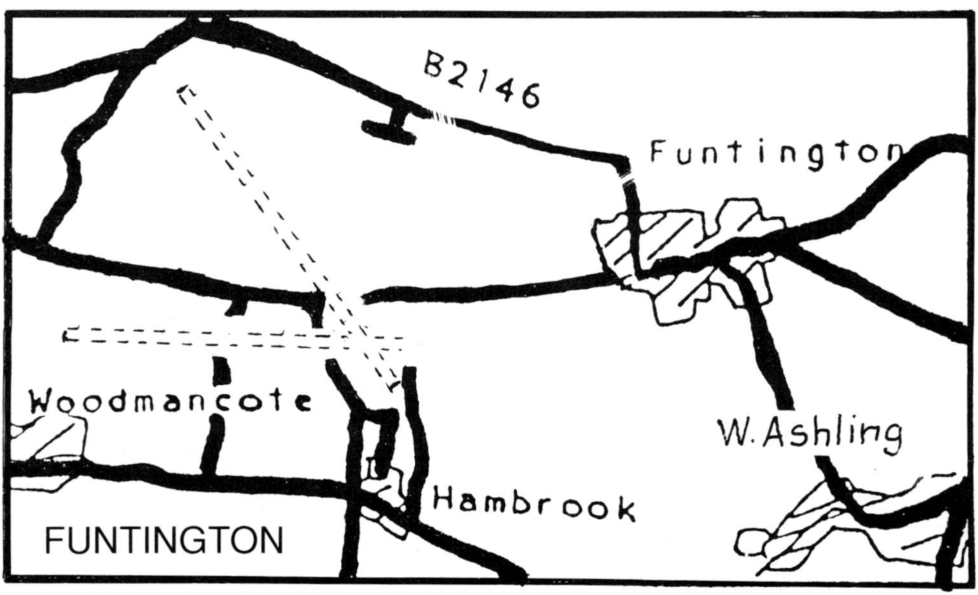

Funtington (West Sussex)

Unit	Arrived from	Date	Left for	Date	Tvpe
2 Sqn RAF	Odiham	15/09/43	Odiham	06/10/43	Mustang I
268 Sqn RAF	Tangmere	15/09/43	Odiham	06/10/43	Mustang I
438 Sqn RCAF	???	01/04/04	Hurn	20/04/44	Typhoon
439 Sqn RCAF	???	01/04/44	Hurn	20/04/44	Typhoon
440 Sqn RCAF	???	01/04/44	Hurn	20/04/44	Typhoon
441 Sqn RCAF	???	02/04/44	???	12/05/44	Spitfire IX
441 Sqn RCAF	???	20/04/44	???	21/05/44	Spitfire XI
443 Sqn RCAF	???	20/04/44	???	12/05/44	Spitfire IX
442 Sqn RCAF	???	02/05/44	???	12/05/44	Spitfire IX
65 Sqn RAF	Ford	14/05/44	Ford	15/06/44	Mustang III
122 Sqn RAF	Ford	14/05/44	Southend	20/05/44	Mustang III
122 Sqn RAF	Southend	28/05/44	Ford	15/06/44	Mustang III
198 Sqn RAF	Thorney lsland	18/06/44	Hurn	22/06/44	Typhoon Ib
609 Sqn RAF	Thorney lsland	18/06/44	Hurn	22/06/44	Typhoon Ib
164 Sqn RAF	Thorneylsland	17/06/44	Hurn	21/06/44	Typhoon Ib
183 Sqn RAF	Thorney Island	18/06/44	Hurn	22/06/44	Typhoon Ib
329 Sqn RAF	Merston	22/06/44	Selsey	01/07/44	Spitfire IX
329 Sqn RAF	Merston	22/06/44	Selsey	01/07/44	Spitfire IX
340 Sqn RAF	Merston	22/06/44	Selsey	01/07/44	Spitfire IX
341 Sqn RAF	Merston	22/06/44	Selsey	01/07/44	Spitfire IX
222 Sqn (Bell)	Coolham	04/07/44	Selsey	06/08/44	Spitfire IX
349 Sqn (Bell)	Coolham	04/07/44	Selsey	06/08/44	Spitfire IX
485 Sqn RNZAF	Coolham	04/07/44	Selsey	06/08/44	Spitfire IX
33 Sqn RAF	Tangmere	12/07/44	Ford	06/08/44	Spitfire IX
66 Sqn RAF	Tangmere	06/08/44	Ford	12/08/44	Spitfire IX
127 Sqn RAF	Tangmere	06/08/44	Ford	12/08/44	Spitfire IX
331 Sqn RAF	Tangmere	06/08/44	Ford	12/08/44	Spitfire IX
332 Sqn RAF	Tangmere	06/08/44	Ford	12/08/44	Spitfire IX

Frost Hill Farm (Hants)

Unit	Arrived from	Date	Left for	Date	Type
Nil	No operational units used this field in 1944				

Headcorn (Kent)

Unit	Arrived from	Date	Left for	Date	Type
403 Sqn RCAF	Lashenden	20/08/43	Kenley	14/10/43	Spitfire XIb
421 Sqn RCAF	Lashenden	20/08/43	Kenley	14/10/43	Spitfire XIb
377 Sqn USAAF	Warningford	13/10/44	Lignerolles	02/07/44	P-47 Thunderbolt
378 Sqn USAAF	Warningford	13/04/44	Lignerolles	02/07/44	P-47 Thunderbolt
379 Sqn USAAF	Warningford	13/04/44	Lignerolles 02/07/44		P-47 Thunderbolt

A Spitfire IX being rearmed with rocket projectiles at an unidentified airfield in Kent.

Rocket armed Typhoon taking off from an un-named advanced landing ground early in June 1944.

surprised when twelve Fortresses and a Mitchell landed there after an operation. However, they coped and regular operations continued. It was in that particular operation that Headcorn's Spitfires shot down seven enemy fighters for the loss of one of their own aircraft.

The RCAF squadrons returned to Kenley for the winter and the site was handed over to the US 9th Air Force, who rebuilt and extended the runways with PSP planking and generally improved the facilities to enable operation of a Wing of Thunderbolts who moved in on 13 April 1944. These operated from Headcorn over northern France, attacking everything in sight to soften up German defences. The Dakotas carrying paratroops on D-Day itself were escorted by Headcorn's Thunderbolts, who then went on to support ground forces as called upon to do so from the ground. This type of mission was continued from the ALG until landing facilities were available in France and Headcorn was released from war service.

High Halden

High Halden was designated as a reserve field when completed in 1943, and because of the ground's reputation of becoming waterlogged it was immediately let out for grazing. It was eventually allocated to the US 9th Air Force who immediately set about improving the facilities before moving in three squadrons of the 358th Fighter Group on 4 April 1944.

At first their main type of mission was to attack railway targets in Northern France to soften up communications prior to the D-Day landings. On D-Day itself they flew escort missions protecting troop carrying formations before they reverted to their previous type of attack in support of ground forces.

When the Americans moved out from High Halden a Flight of Meteors from 616 Squadron RAF were flown in from Manston to operate 'Anti Diver' patrols which involved intercepting and, with luck, shooting down V-1 Flying Bombs before they reached their targets in the London area. Apart from a Spitfire that crash landed, these 'Anti Diver' operations were the last from High Halden and the site was derequisitioned in September 1944 when the American engineers were allowed to remove the PSP for re-use on the Continent.

High Halden (Kent)

Unit	Arrived from	Date	Left for	Date	Type
365 Sqn USAAF	Raydon	13/04/44	Cretteville	13/07/44	P-47 Thunderbolt
366 Sqn USAAF	Raydon	13/04/44	Cretteville	03/07/44	P-47 Thunderbolt
367 Sqn USAAF	Raydon	13/04/44	Cretteville	03/07/44	P-47 Thunderbolt
616 Sqn RAF (1 Flt)	Manston	??/08/44	Manston	??/09/44	Meteor I

Larks Barrow (Hants)

Although surveyed in 1942 as a second line ALG, there is no evidence that apart from tree lopping and general site clearance any construction ever took place. It is thought that the only aviation use the site has been put to was for forced landing training by aircraft from nearby units.

Kingsnorth

This ALG should not be confused with the more famous field of the same name used by the Admiralty as an airship station in the First World War; that was north-east of Rochester. The World War 2 Kingsnorth ALG was three miles southeast of Ashford, one of a group of five ALGs all quite close to one another.

When this site was completed in 1943 a number of units were moved in with the dual purpose of carrying out necessary operations and at the same time learning to work from crude and minimal airstrips. This was so that they could learn what to expect when flying close support missions during the early stages of the projected invasion of Europe.

After a number of improvements carried out during the winter months of 1943-44, three squadrons of the 36th Fighter Group, 303rd Flying Wing of the US XIX Tactical Air Force flew in their Thunderbolts and immediately started the mixture of operations all aimed at softening the enemy's defences prior to D-Day. They flew beachhead patrols in the initial stages of the invasion, and later attacked missions in ground support until they were able to fly from captured sites in France in July 1944. Two months later the Kingsnorth site was derequisitioned and the PSP lifted for further use on the Continent, handing back the land to its previous farming owners.

Kingsnorth (Kent)

Unit	Arrived from	Date	Left for	Date	Type
65 Sqn RAF	Selsey	01/07/43	Ashford	05/10/43	Spitfire Vb
122 Sqn RAF	Bognor	01/07/43	Brenzett	14/09/43	Spitfire Vb
602 Sqn RAF	Bognor	01/07/43	Newchurch	13/08/43	Spitfire Vd
19 Sqn RAF	Newchurch	18/08/43	Weston Zoyland	30/09/43	Spitfire IX
184 Sqn RAF	Manston	14/08/43	Newchurch	17/09/43	Hurricane II
22 FS USAAF	???	04/04/44	Bruchville	17/07/44	P-47 Thunderbolt
23 FS USAAF	???	04/04/44	Bruchville	17/07/44	P-47 Thunderbolt
53 FS USAAF	???	04/04/44	Bruchville	17/07/44	P-47 Thunderbolt

Lashenden (Kent)

Unit	Arrived from	Date	Left for	Date	Type
403 Sqn RCAF	Kenley	06/08/43	Headcorn	18/08/43	Spitfire
421 Sqn RCAF	Kenley	06/08/43	Headcorn	18/08/43	Spitfire
353 Sqn USAAF	Boxted	17/04/44	Criqueville	23/06/44	Mustang P-51 B
355 Sqn USAAF	Boxted	15/04/44	Criqueville	23/06/44	Mustang P-51 B
356 Sqn USAAF	Boxted	17/04/44	Criqueville	23/06/44	Mustang P-51 B

Author and publisher would welcome any information that completes unknown dates and locations indicated in the tables.

Lashenden

Before being approved as an ALG, Lashenden had previously been proposed as a fighter bomber base. However, it was eventually decided that together with Lydd it would be a training base for the mobile squadrons being formed for the invasion of Europe. The first squadrons to arrive were Spitfires of the RCAF who carried out a few operations which included shooting down one Me 109. After only twelve days they moved on to Headcorn and the site was stood down for a number of improvements to be made, including the by then mandatory increase of hardstanding areas, increase of taxiways and extension of runways.

In April 1944 it was handed over to 100th Fighter Wing of the XIX Tactical Air Command with their Merlin-engined Mustangs. Their main tasks were the long-range escort of Flying Fortresses and Mitchell bombers. This wing was able to claim real success in this unenvied task as the record of 20 enemy aircraft destroyed for the loss of only two Mustangs shows.

After a month, fighter bomber missions were mounted from Lashenden and later the Mustangs were designated as escorts to glider formations on the afternoon of D-Day itself. Other operations carried out before leaving Lashenden by these versatile formations was close support work on the battle front and 'Anti Diver' (V-l) patrols before leaving for France later in June of 1944. By October 1944 work was started to strip the warlike equipment from the site and return it to agriculture in response to great pressure from the Ministry of Agriculture.

A quarter of a century later aviation returned to Lashenden when a corner of the old ALG was used by civil light aircraft. Today it is being operated by Shenley Farms (Aviation) Ltd as a private airfield and Flying School under the confusing name for airfield historians of Headcorn!

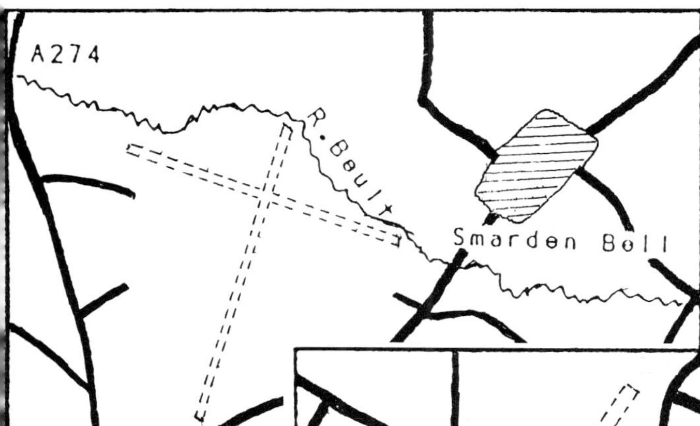

LASHENDEN

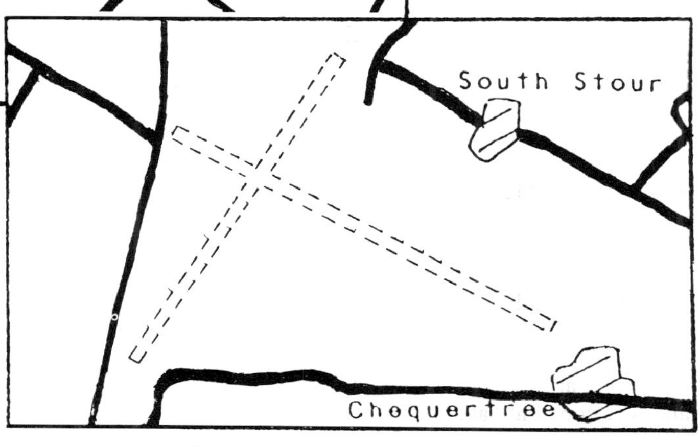

KINGSNORTH

Lydd

Lydd had been connected with aviation in the form of balloons since 1886 and later with various activities of the RFC until relinquished in 1918. Its potential as an ALG was realised in World War 2 and Sommerfield track runways were laid after much site drainage had been affected.

In July 1943 three Squadrons of Typhoons arrived from Merston, Appledram and Selsey to learn to work as a Wing. Various aggressive operations were carried out with considerable success and Lydd itself received an air raid. There was little damage and no casualties to interrupt the programme of dive bombing and offensive sweeps across the Channel. The heavy fighter-bomber version of the Typhoon played havoc with the Sommerfield tracking and actually caused a few accidents among those flown from Lydd. The wing was moved to Westhampnett (Goodwood) for the winter while repairs and improvements were made to the airfield.

In fact, it was never used again operationally, being used to house various detachments of the RAF Regiment throughout the Summer of 1944. By the September of that year no further use for the site was considered and it was derequisitioned and cleared early in 1945.

Lymington

Another advanced airfield was built to cope with the build-up for D-Day at Snooks Farm, about a mile to the east of Lymington.

The official name for this airfield was Lymington, but many of the locals refer to it as Pylewell after the nearby estate, or sometimes Wallhampton. As with the other ALGs the expected life of this airfield was only planned to be a matter of months and facilities were confined to the minimum required to sustain operations for that sort of time period. Two temporary runways were laid down using steel wire mesh to reinforce the frequently soft low lying ground overlooking the Solent. The facilities for equipment fared better than personnel with several blister type hangers being erected for stores and limited maintenance facilities. Humans and aircraft had to be content with the open air or tents.

During March 1944 the American 50th Fighter Group from IX Tactical Air Command moved in with three Squadrons, Nos 10, 81 and 313, all with P-47 Thunderbolts to start intensive training for the operations to come. Their first mission was on May 1st when they

Lydd (Kent)

Unit	Arrived from	Date	Left for	Date	Type
172 Sqn RAF	Merston	01/07/43	West-hampnett	10/10/43	Typhoon IB
175 Sqn RAF	Appledram	02/07/43	West-hampnett	09/10/43	Typhoon IB
245 Sqn RAF	Selsey	30/06/43	West-hampnett	10/10/43	Typhoon IB

Lymington (Hants)

Unit	Arrived from	Date	Left for	Date	Type
10 Sqn USAAF		??/03/44	France	25/04/44	P-47 Thunderbolt
81 Sqn USAAF		??/03/44	France	25/04/44	P-47 Thunderbolt
313 Sqn USAAF		??/03/44	France	25/04/44	P-47 Thunderbolt

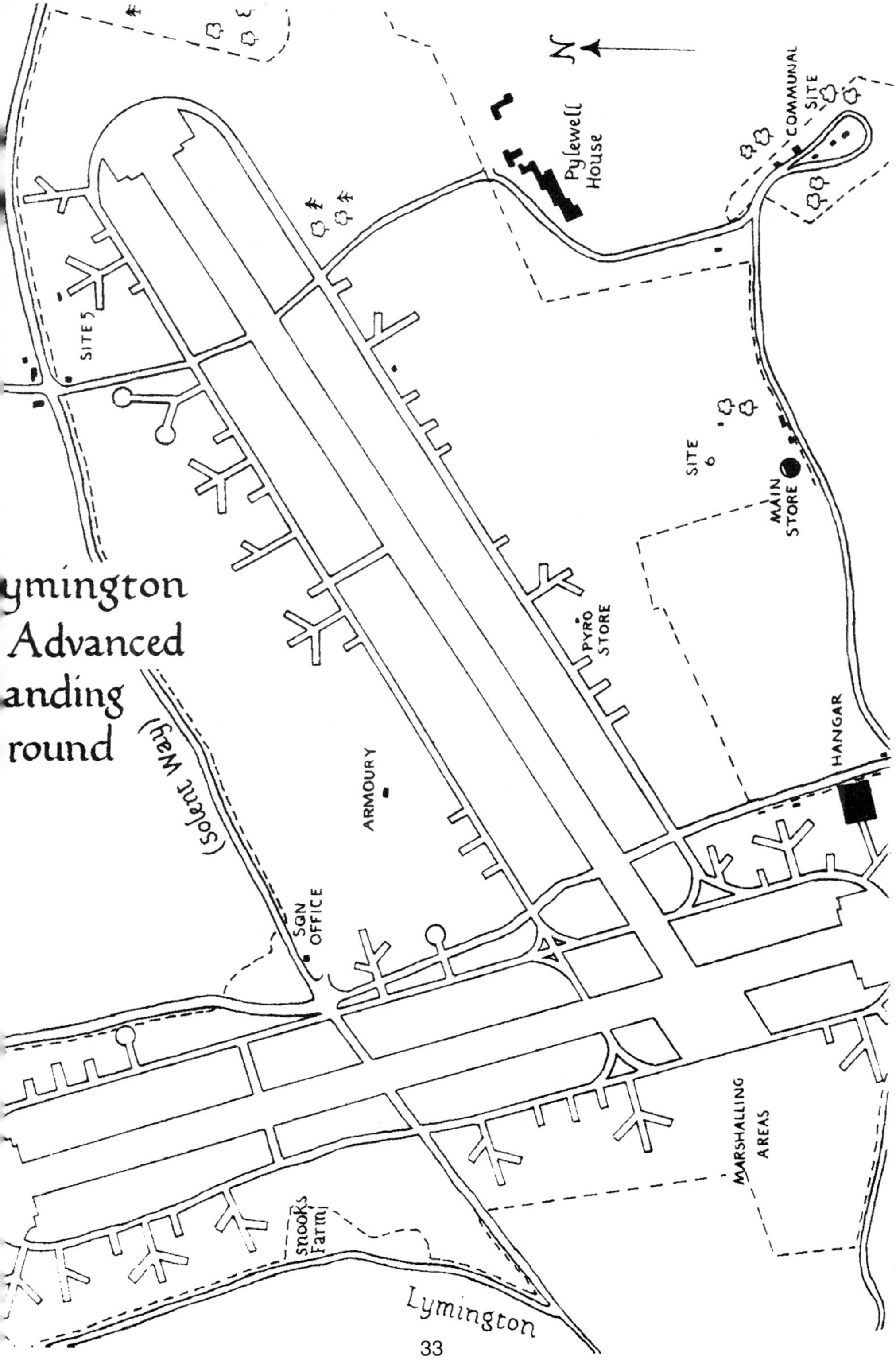
ymington
Advanced
anding
round
N
Pylewell House
COMMUNAL SITE
SITE 5
SITE 6
MAIN STORE
PYRO STORE
HANGAR
ARMOURY
(Solent Way)
SQN OFFICE
MARSHALLING AREAS
Snooks Farm
Lymington

started their allotted task of softening up communications in Northern France.

A few weeks after D-Day landing strips became available in France and all aircraft based at Lymington moved across to the Continent. leaving the airstrip that had served them for less than four months with its task complete. With no further use for the site, it was soon derequisitioned and allowed to return to agriculture.

Needs Ore Point

The New Forest's share of ALGs were built at the four different sites and work started on the first to be established as an advanced airfield at Needs Ore Point in the middle of 1943. The situation was on the low lying land just to the west of the entrance of the Beulieu River screened from the north by trees, which later proved useful for hiding the aircraft. As an advanced airfield it could hardly have been more advanced as far as position was concerned, but certainly not in facilities which were crude to say the least. The usual steel wire mesh was laid to form runways, accommodation for everything and everyone was in tents, and only facilities for first line servicing were available on site.

One development in 1943 that involved flying over Needs Ore Point and which puzzled the few local residents was the erection of what seemed to be a massive hoarding or screen. Nobody could find out what it was and enquiries today still haven't revealed what it was used for.

One explanation put forward suggested it was a target for radio-controlled aircraft which were frequently seen in the area, but a local resident maintains that his father helped to build it for secret exercises given to the SOE who underwent some of their training at nearby Beaulieu. Although this makes an interesting sidelight on possible aviation activities in the Forest, it certainly had no connection with the advanced landing ground at Needs Ore Point.

Between December 1943 and April 1944 some extra work was carried out to ready the field to receive 146 and 149 Squadrons with their Typhoons from Tangmere. One day after they arrived 193 Squadron flew in with their Typhoons and a day or so later the assigned units reached full strength when they were joined by 257 and 266 Squadrons. The immensity of this operation from this tiny and temporary airstrip cannot be described in words alone, but the following figures were very revealing.

Like other group formations being worked up for this phase of the war, the two at Needs Ore, Nos 83 and 84, were set up to be fully mobile. They were supported by 900 men, not including pilots, 200 lorries and each of the constituent squadrons operated at least 30 aircraft. At its very short peak of activity, Needs Ore Point had 150 aircraft based there and over one hundred of them were kept operational at any one time.

During the run up to D-Day and immediately afterwards operations from Needs Ore were so frequent there probably isn't an airfield in the British Isles today which can boast the intensity of flying that took place from this ALG for a few weeks in the summer of 1944. When researching this material, I was lucky enough to meet a pilot who actually flew operations from Needs Ore Point and who told me that on four fine days prior to D-Day there was a take-off and landing at this strip every 45 seconds during daylight hours, and this without modern radar facilities.

Needs Ore Point (Hants)

Unit	Arrived from	Date	Left for	Date	Type
257 Sqn RAF	Tangmere	10/04/44	Hurn	02/??/44	Typhoon IB
266 Sqn RAF	Tangmere	10/04/44	Snaith	27/06/44	Typhoon IB
266 Sqn RAF	Snaith	06/05/44	Eastchurch	29/06/44	Typhoon IB
197 Sqn RAF	Tangmere	10/04/44	Hurn	03/07/44	Typhoon IB
193 SqnRAF	Llanbeder	11/04/44	Hurn	03/07/44	Typhoon IB

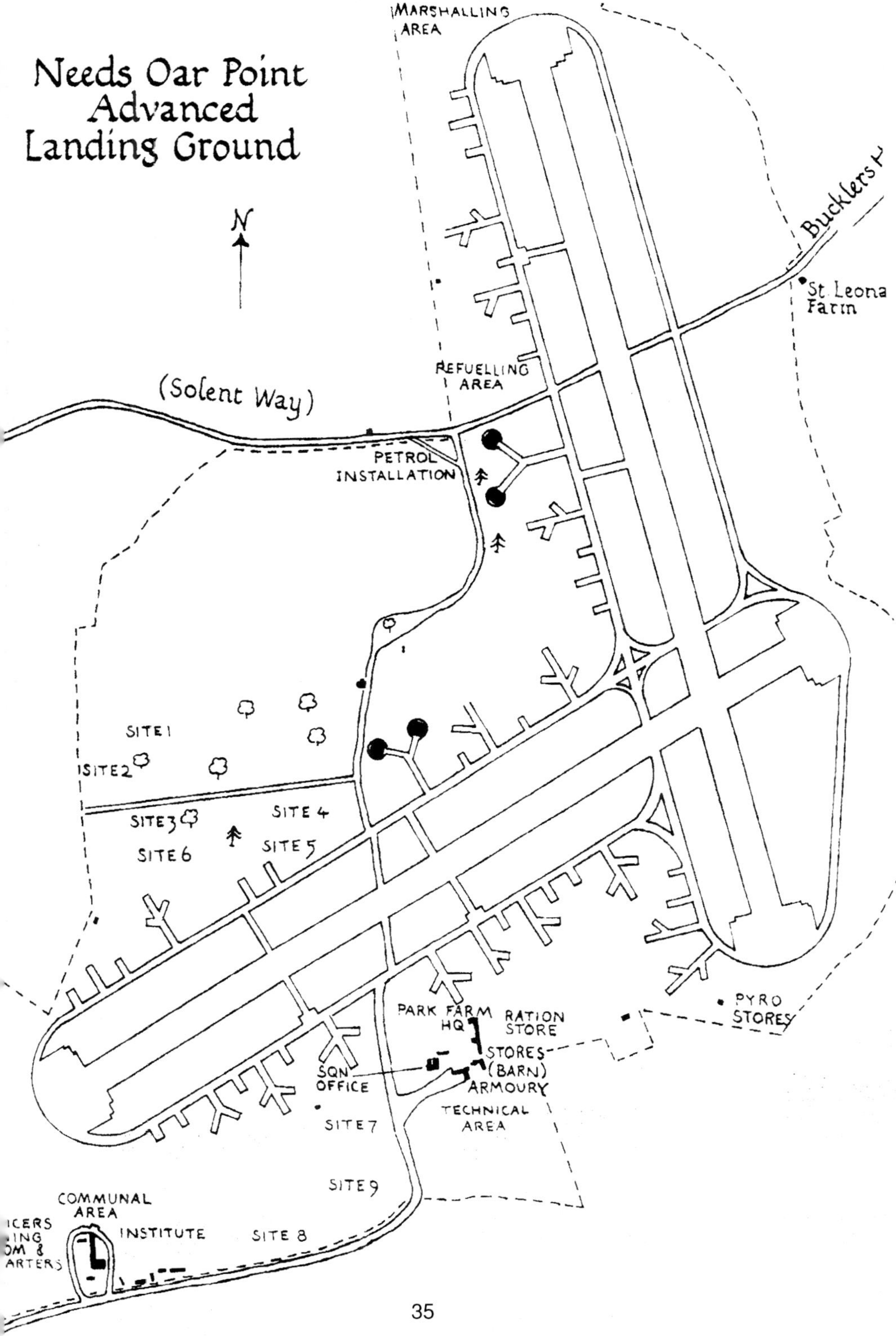
Needs Oar Point Advanced Landing Ground
N
MARSHALLING AREA
Bucklers H
St. Leona Farm
(Solent Way)
REFUELLING AREA
PETROL INSTALLATION
SITE 1
SITE 2
SITE 3
SITE 4
SITE 5
SITE 6
SITE 7
SITE 8
SITE 9
PARK FARM HQ
RATION STORE
STORES (BARN)
ARMOURY
SQN OFFICE
TECHNICAL AREA
PYRO STORES
COMMUNAL AREA
INSTITUTE
ICERS ING OM & ARTERS

As soon as airfields were available in France, Needs Ore became redundant and deserted, although the airfield itself was left open for a few months to serve as an emergency landing ground. The Royal Navy used it for storage purposes until 1946 when it was derequisitioned and the land returned to agriculture, marking the end of its brief but spectacular existence as one of the New Forest's airfields. Forty-five years later the views are there, but apart from some gaps in the hedgerows no sign of an airfield.

Newchurch

By agreement with the Ministry of Agriculture, Newchurch was to be handed back for grazing until it was needed for operations as soon as construction work was completed. This arrangement didn't last for long as two squadrons of Spitfires were moved in on July 2 1943 to provide escorts for bombers operating from East Anglian bases. After some three months of hectic and some not altogether successful operations, the units based there were withdrawn to Detling to enable improvement in the facilities to be made over the winter period.

A strange mixture of Spitfires, Tempests and Typhoons moved in during April 1944 due to late delivery of the expected Tempests, and operations on shipping 'recce' flights were commenced.

The various dull, but dangerous missions that fell to the lot of Newchurch based squadrons was rewarded when a couple of days after D-Day No 3 Squadron was guided by controllers onto a formation of Me 109s. After a short but hectic fight four enemy aircraft were shot down and several more severely damaged without loss to the Newchurch Tempests.

The aircraft based here were also very much involved with 'Anti Diver' (V-l) operations, a job at which they got very good after many experiments with attack techniques and some illegal alteration to their armament. One of the problems they had to face was not their aircraft, but the noise of the nearby anti-aircraft guns, and the shrapnel from them falling on the tented accommodation. By September 1944 things had become a little quieter and the Wings were moved across to France or to another base for a well deserved rest.

Newchurch was one of the more successful of the ALGs and derequisitioned in December 1944, leaving nothing to show today of the battles that were fought from there.

Newchurch (Kent)

Unit	Arrived from	Date	Left for	Date	Type
19 Sqn RAF	Bognor	02/07/43	Kingsnorth	18/08/43	Spitfire Vb
132 Sqn RAF	Gravesend	02/07/43	Detling	12/10/43	Spitfire Vb
602 Sqn RAF	Kingsnorth	02/07/43	Detling	12/10/43	Spitfire Vb
184 Sqn RAF	Snailwell	17/09/43	Detling	12/10/43	Hurricane IId
3 Sqn RAF	Bradwell Bay	28/04/44	Matlask	21/09/44	Tempest V
56 Sqn RAF	Ayr	28/04/44	Matlask	23/09/44	Spitfire IX & Typhoons
486 Sqn RNZAF	???	01/05/44	???	21/09/44	

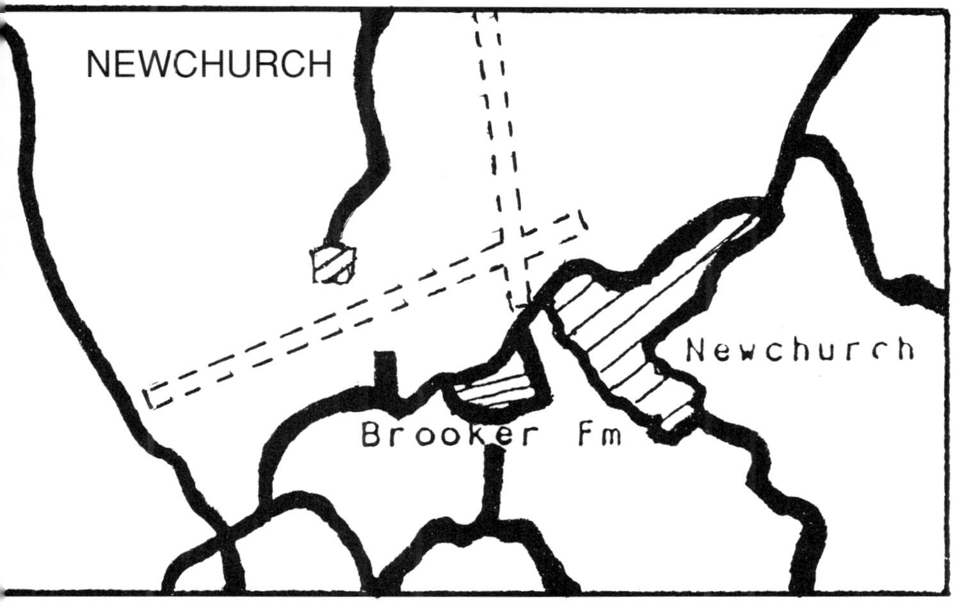

A Spitfire IX with its invasion force identification stripes being re-armed in a quiet corner of an advanced landing ground, probably Chailey in East Sussex.

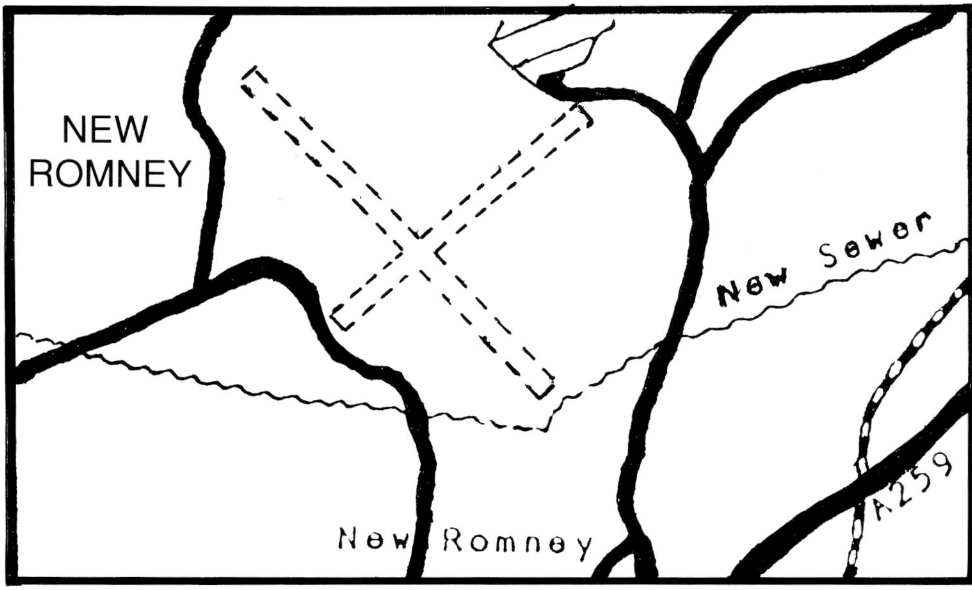

New Romney

When approved it was realised that New Romney was a very exposed site and there was little or no possibility of any camouflage that would disguise the field. A decoy site was built a mile or so away and Typhoon fighters and fighter bombers moved in to take part in various operations as part of a practice programme for future operations from forward airfields. Honeychild Manor, a large shed, some old army huts left behind by a searchlight unit and four cottages provided what for ALGs, was good accommodation.

In spite of further improvements carried out during the winter months New Romney remained a reserve field during the invasion and only used to accommodate various signal units waiting to move across the Channel after D-Day. Apart from a damaged Liberator that crash landed there in July even its declared assignation as an emergency landing ground was little used and the site was handed back to its owners for agriculture in December 1944 after a fire in the Manor House.

New Romney (Honeychild) (Kent)

Unit	Arrived from	Date	Left for	Date	Type
181 Sqn RAF	Appledram	02/07/43	Merston	08/10/43	Typhoon
182 Sqn RAF	Appledram	02/07/43	Wigtown	18/09/43	Typhoon
247 Sqn RAF	Bradwell Bay	10/07/43	Attlebridge	07/08/43	Typhoon
247 Sqn RAF	Attlebridge	13/08/43	Merston	11/10/43	Typhoon

Selsey

Selsey started its aviation life as a private airfield next to Pagham Harbour, but in spite of strong opposition from the Ministry of Agriculture was designated as a priority site for the construction of an ALG. This was because of its clear approaches, flat surface and reasonable accommodation in the area. One or two of the larger houses and an old hangar from the private flying days were retained.

A Spitfire squadron arrived from Fairlop at the end of May 1943 followed by a Typhoon squadron the next day. Training missions were started straight away, the Spitfires carrying out escort duties and the Typhoons practicing ground attack techniques. The winter months saw extra blister hangars and further hardstanding areas added in preparation for the expected activities that would be necessary for invasion operations.

In April 1944 Spitfire IXs arrived and immediately became involved in dive bombing and escort missions aimed at softening up communications on the Continent. They provided the much needed air cover over the beaches on D-Day, shooting down a number of enemy bombers and other aircraft while doing so. After D-Day a host of different types of operations were carried out from Selsey, all vital in support of the troops advancing across Northern France. Four and a half months after the first squadrons arrived in April, the Wing moved a few miles northwards to Tangmere leaving the site to standby as a reserve field until March 1945 when it was returned to agriculture. In spite of this the Air Ministry kept their hands on it until after the war was over when the runways were lifted, hangars removed and Selsey finally derequisitioned.

Considering the great number of operations mounted from this crude airfield, it has to be considered as one of the more successful of the ALGs.

Selsey (West Sussex)

Unit	Arrived from	Date	Left for	Date	Type
65 Sqn RAF	Fairlop	31/05/43	Kingsnorth	01/07/43	Spitfire V
245 Sqn RAF	Fairlop	01/06/43	Lydd	30/06/43	Typhoon Ib
485 Sqn RNZAF		??/??/??		??/??/??	
222 Sqn RAF	Southend	11/04/44	Coolham	30/06/44	Spitfire IX
349 Sqn RAF	Hornchurch	11/04/44	Coolham	30/06/44	Spitfire IX
340 Sqn RAF	Funtington	01/07/44	Tangmere	14/08/44	Spitfire IX
341 Sqn RAF	Funtington	01/07/44	Tangmere	06/08/44	Spitfire IX
74 Sqn RAF	Tangmere	17/07/44	Southend	24/07/44	Spitfire IX
33 Sqn RAF	Funtington	06/08/44	Fairwood Common	12/08/44	Spitfire IX
33 Sqn RAF	Fairwood Common	18/08/44	B10 Plumetot	19/08/44	Spitfire IX
222 Sqn RAF	Funtington	06/08/44	Tangmere	19/08/44	Spitfire IX
349 Sqn RAF	Funtington	06/05/44	Tangmere	19/08/44	Spitfire IX

160 - C4

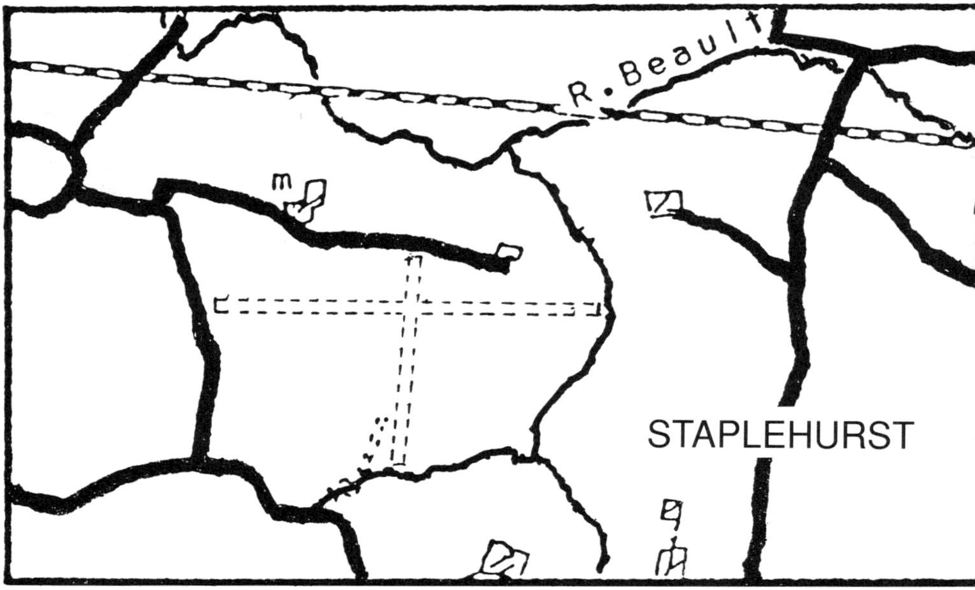

Staplehurst

The most westerly of the Kentish Advanced Landing Grounds, Staplehurst, was completed in March 1943 and handed over for grazing until a Canadian Wing moved in to take part in various escort missions and fighter sweeps over France. They moved to their winter quarters in October, the airfield being put through the programme of improvements derived from the lessons learned from operations there and at other ALGs during 1943.

Three squadrons of P-51Ds of the 363rd Fighter Group of US IX Tactical Air Command moved in to take over escort duties to heavy bombers. At the end of May they were placed on offensive missions against communications, reverting to escorts for glider formations on D-Day. On completion of this one-way trip for the gliders, and before returning to Staplehurst, they strafed enemy positions.

The Americans left for France on July 1, 1944, and the airfield was derequisitioned in October, finally being released for agriculture in January, 1945.

Staplehurst

Unit	Arrived from	Date	Left for	Date	Type
401 Sqn RCAF	???	06/08/43	Biggin Hill	14/10/43	Spitfire Vb
411 SqnRCAF	???	06/08/43	BigginHill	14/10/43	Spitfire Vb
412 Sqn RCAF	???	06/08/43	Biggin Hill	14/10/43	Spitfire Vb
380 Sqn 363FG USAAF	Rivenhall	01/04/44	Maupertuis	01/07/44	P-51 D Mustang
381 Sqn 363FG USAAF	Rivenhall	01/04/44	Maupertius	01/07/44	P-51D Mustang
382 Sqn 363FG USAAF	Rivenhall	01/04/44	Maupertuis	01/07/44	P-51 D Mustang

Swingfield

Swingfield has aviation connections going back to 1916 when it was designated an unmanned emergency landing ground. It was again used, by Swordfish in 1940, when they flew about as decoys pretending to be Gladiator fighters.

The site was surveyed and an almost standard Advanced Landing Ground was constructed in 1943. A few Spitfires landed and took off again to test the site out and having found it satisfactory the Ministry allowed it to be returned to the farmers for grazing.

It was never used in the D-Day operations and apart from a few aircraft from Hawkinge and alleged use by SOE Lysanders, it remained unused until in August, 1944 when ground 'Anti Diver' defences moved in asking the Swordfish aircraft they found there to remove themselves back to their base at Inskip. Eleven Albacores from 119 Squadron RAF arrived a few days after the 'Stringbags' left and were joined later by more from 819 Sqn FAA. They used the base for anti-shipping patrols in the Channel until they all flew off to Bircham Newton in October. The site was finally derequisitioned in April, 1945 when the hangars and runways were removed.

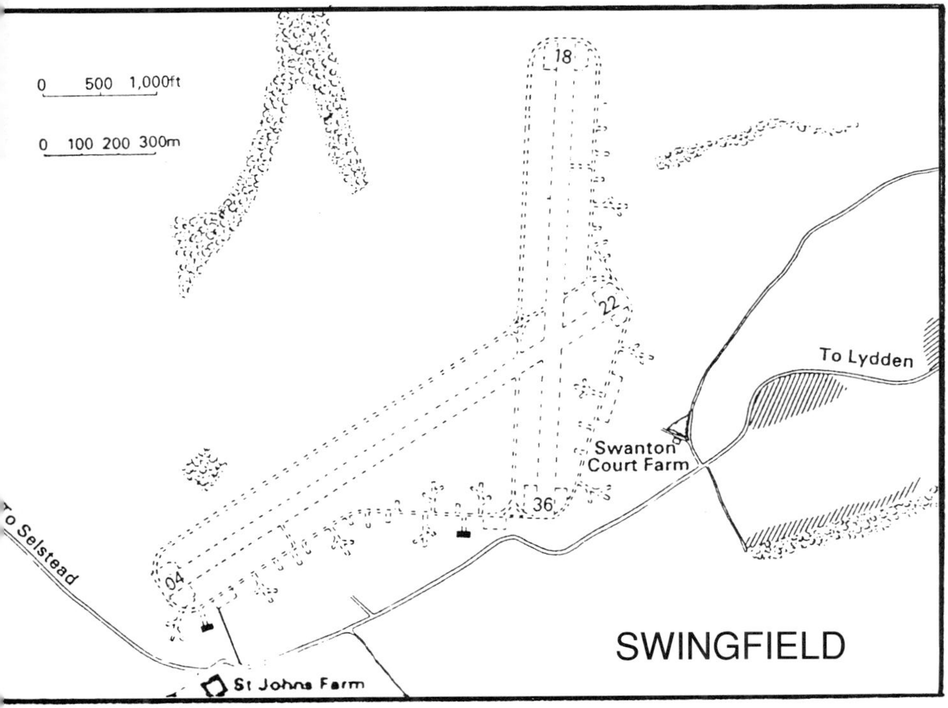

Swingfield (Kent)

Unit	Arrived from	Date	Left for	Date	Type
119 Sqn RAF	Manston	09/08/44	Bircham Newton	02/10/44	Albacore I
819 Sqn FAA	???	??/08/44	Bircham Newton	02/10/44	Albacore I

Winkton

The site for the third New Forest ALG was selected at Winkton, frequently referred to by the locals as Sopley. In the same way as the other advanced landing grounds the land was cleared and levelled where necessary. Two wire mesh runways were laid down, and like others of its type this airfield was only set up with the absolute minimum of facilities and recorded as being ready by September, 1943. The contractors had done well because as it happened it wasn't needed straight away and it wasn't until April of the next year that the American 404th Fighter Group consisting of Nos 506, 507 and 508 Squadrons with their P-47 Thunderbolts arrived.

These aircraft were operated as fighter bombers and concentrated on locating and destroying where possible the V-1 sites. On D-Day itself the whole of 404 Group was providing top cover for the landings taking place on the beachheads when they flew a total of 191 missions from Winkton between dawn and dusk.

Many people to this very day confuse RAF Sopley with the advanced landing ground built on an adjacent site known as Winkton. This is understandable because those who are not familiar with the events of 1944 are probably only aware of the School of Air Traffic Control which was until quite recently at Sopley. Like most of the other Advanced Landing Grounds, Winkton served its purpose in a few hectic months and the site was relinquished as soon as operations allowed the aircraft to move over to the Continent.

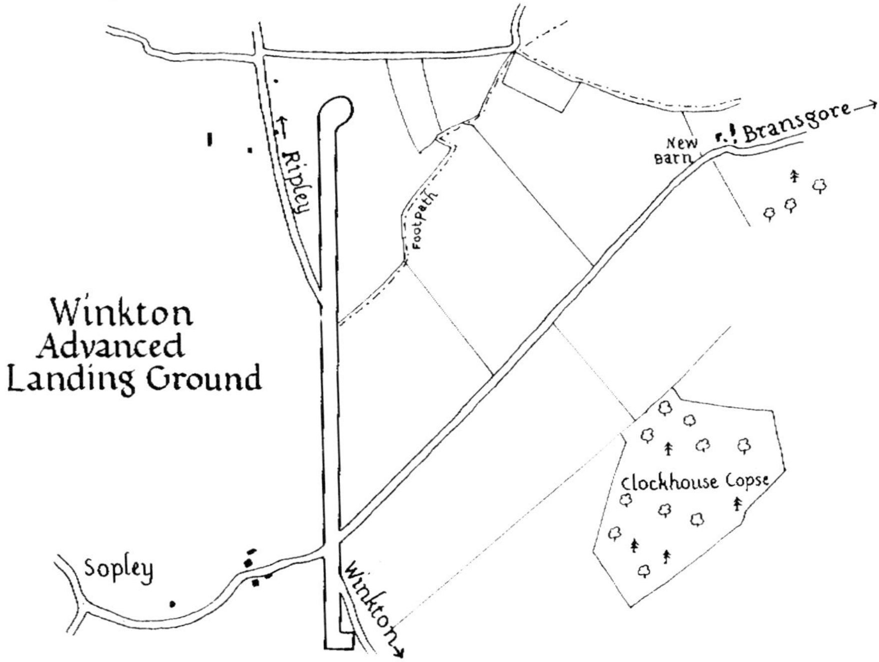

Winkton (Hants)

Unit	Arrived from	Date	Left for	Date	Type
507 Sqn USAF	???	??/03/44	France	10/06/44	P-47 Thunderbolt
508 Sqn USAF	???	??/03/44	France	10/06/44	P-47 Thunderbolt

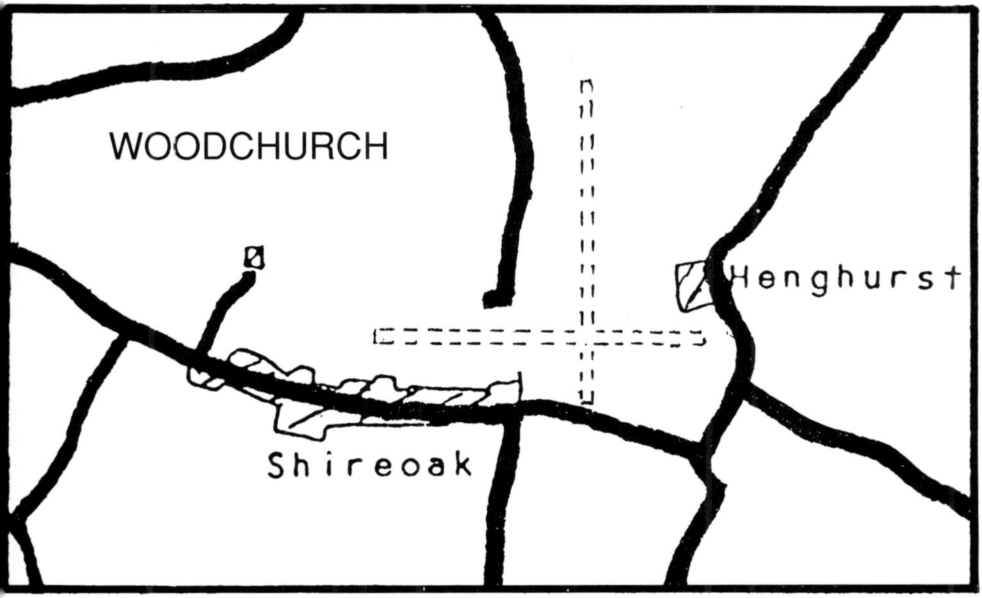

Woodchurch

At one time Woodchurch was destined to be an advanced base for light bomber operations, but after difficulties with the actual site it was regulated to ALG status and two squadrons of RCAF Mustangs were flown in from Dunsfold at the end of July, 1943. In the two and a half months they stayed at Woodchurch not only did these two units carry out essential photo recce flights over the Channel, but did very well on train busting missions and generally creating mayhem all over Northern France. The Canadians moved to winter quarters at Redhill leaving the winter months for the airfield to be prepared for three squadrons of P-47 Thunderbolts of 373rd Fighter Group, IX Tactical Air Command USAAF, to take over on 4 April, 1944.

This Wing provided escort cover for medium bombers softening up communication targets, top cover over the beachheads on D-Day, followed by as many aggressive missions as they could fit into time available. They moved to the Continent on 4 July, 1944.

Woodchurch (Kent)

Unit	Arrived from	Date	Left for	Date	Type
231 Sqn RCAF	Dunsfold	28/07/43	Redhill	15/10/43	Mustang I
400 Sqn RCAF	Dunsfold	28/07/43	Redhill	15/10/43	Mustang I
410 FS USAAF	???	04/04/44	Touen Bassin	04/07/44	P-47 Thunderbolt
411 FS USAAF	???	04/04/44	Touen Bassin	04/07/44	P-47 Thunderbolt
412 FS USAAF	???	04/04/44	Touen Bassin	04/07/44	P-47 Thunderbolt

LANDING GROUNDS USED F

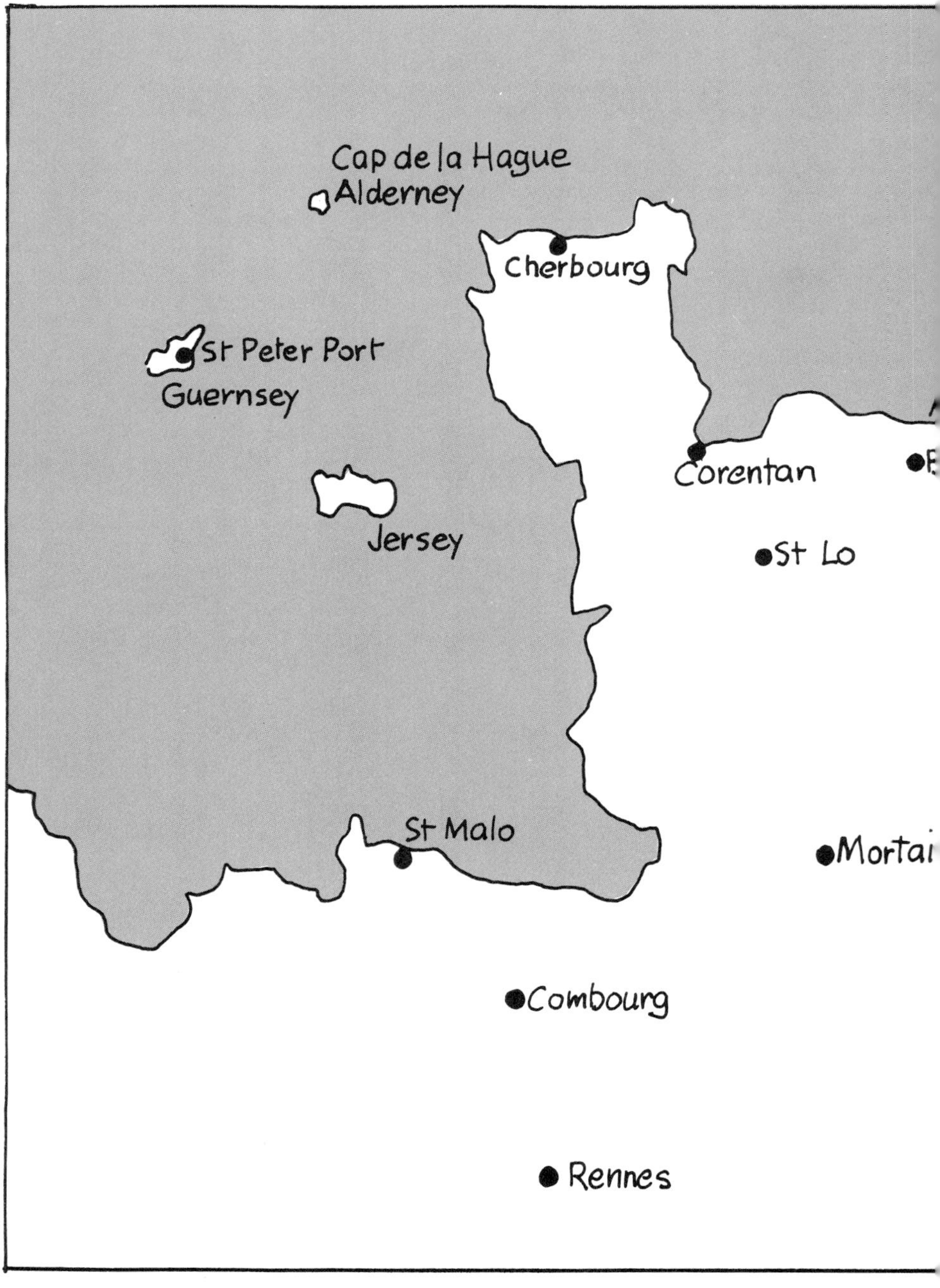

LLIED FORCES AFTER 'D' DAY

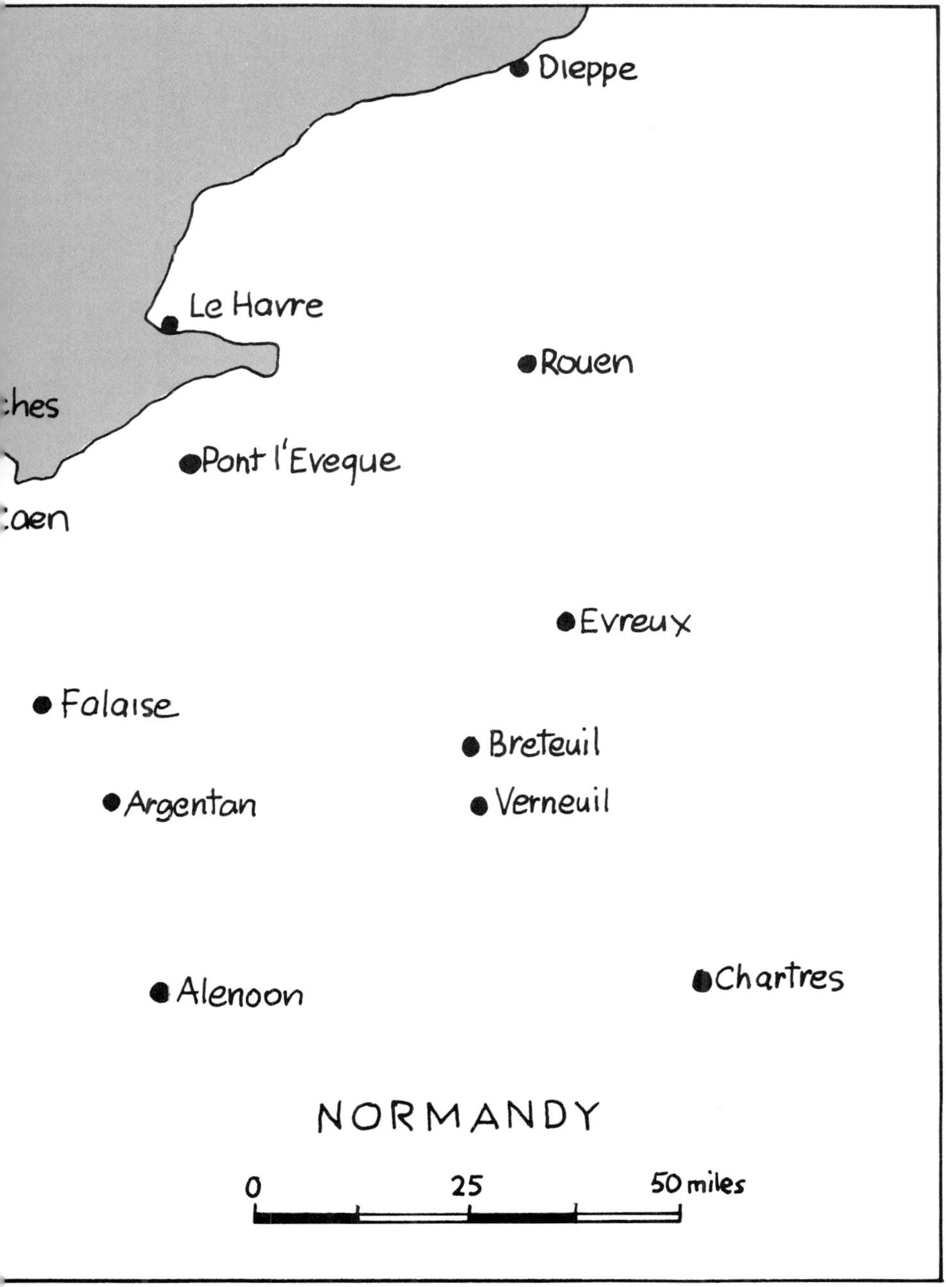

Ninety Days Plus

As will be seen from the various tables listing the units flown from the Advanced Landing Grounds, the vast majority of them were not truly operational until April, 1944, and in less than a month after the D-Day landings were deserted.

They were three main reasons for all this effort being wound up so suddenly after only three months' use. The prime reason was that the construction of each of the sites was only intended to provide a temporary advanced airfield as near as possible in Britain to the coast of France where the landings were to take place. The Ministry of Agriculture, who had opposed the takeover of good farm land in the first instance, made strong claims to the land once there was no absolute operational need for the sites. Second, once a foothold in Normandy was established close support could be given to ground forces by using captured sites across the Channel. And finally, chronic shortage of steel mesh meant that all available material was needed for the establishment of runways in France. This involved a massive logistic organisation just to lift the mesh from the ALG in Britain, transport it across the Channel and reassemble it on suitably cleared fields in Northern France. It was to the credit of the engineering battalions that this was achieved in the midst of one of the largest battles the world has ever seen.

Although it is 46 years since D-Day, many of the old ALGs are remembered by those who once flew from them. The majority are nearly impossible to recognise today, making it difficult to imagine the fiendish activity that took place for a few months from these now peaceful fields. In one case, that of Christchurch, even the aircraft factory has gone, a new industrial estate has sprung up and houses now cover much of the site. Among local people there are many with memories of the contribution to aviation in general and the war effort in particular made by Christchurch. Others have only the vague indication in the names of some roads around there today such as Airspeed Road, De Havilland Way, Comet Way, and The Runway, indicating that once gliders, training aircraft, fighters British and American jets and airliners all flew in and out of the old airfield.

One or two of the old sites have recognisable buildings left in situ which in some cases like the old blister hangar at Lymington has been found useful for the storage of farm equipment. Lydd ALG also has an old Nissen hut which is said to have served the ALG as OC's office.

Other buildings that have withstood the ravages of time in one or two places often present a complete mystery as to their original purpose. Memory plays funny tricks after nearly half a century and even talking to those who served at the time can elicit two different answers to the same question.

This is the case with the simple rectangular brick buildings with concrete roofs which resemble the old civilian street air raid shelters and can be seen at Coolham and Chailey. Nobody seems to remember what they were; it may be that they were air raid shelters, or more likely ammunition stores. Perhaps one day we will find out.

A few of the old ALGs have small lengths of taxiways and portions of hardstandings left in situ. Frequently where this is so the farmers put them to good use or, as in the cases of Deanland, the remains are used for roadways and parking places on caravan sites.

Although not actually ALGs, Holmsley South, Stoney Cross and the old Beaulieu airfields in the New Forest are good examples of this type of later use of an old airfield. Those interested in delving deeper into the exploration of ALGs might find a visit to these more accessible sites rewarding in order to obtain a better idea of what to look for.

Within ten miles of these New Forest sites were four ALGs, Needs Ore Point, Lymington, Winkton and Bisterne. They are typical of ALGs that were returned to agriculture when their use as an airfield was over and also typical of how most of the others are now, 45 years later. Apart from a few odd buildings most are now only identifiable by gaps in hedgerows with fencing marking where the temporary runways used to traverse the old field boundaries.

Other signs are where the landowners have put to use such remains of the steel mesh used for the runways that wasn't removed when the Air Forces moved out.

This brings us to a very important point. The vast majority of these old sites are now private property and/or in use for farming. For example, Needs Ore Point, part of the Beaulieu Estate, is now a nature reserve. This makes it absolutely essential that anyone wishing to explore old ALGs should seek permission first. This applies to all those sites on private property. If approached properly permission is seldom refused, but even so consider this the most important and first of the 'Aviation Archaeologist's Ten Commandments'.

As has been mentioned before on these pages, life is made easy for us at Appledram where a D-Day airfield museum is situated and for a very modest sum the whole family can see for themselves what life was like on an ALG.

At the other end of the scale the old Lashenden has been renamed Headcorn and part of the old Advanced Landing Ground used as a private civil and licenced airfield. Although returned to Agriculture in 1945, Shenley Farms (Aviation) Ltd reopened the western end for use by light aircraft. Weald Air Services Ltd operated air taxi, charter services and pleasure flights from 1973 onwards.

At one point the Maidstone and Ashford Borough Councils ordered its closure, but a licence was obtained for operation of aircraft up to 5,670kg and the closure order was rescinded. It now has two grass runways and is becoming increasingly popular, with the Lashenden Air Warfare Museum, a thriving Microlight club, Weald Air Services, a parachute club, and a fair number of privately owned light aircraft based there, all adding up to a considerable scene of aviation on this old Advanced Landing Ground.

Some of the 23 old ALGs are all but lost, leaving the enthusiast and archaeologist with a lot to explore, and there are only one or two with much to show of their brief glory in the 'Ninety Days to Normandy'.

Typoon of 175 Squadron about to be scrambled from Lydd on a sweep across Northern France immediately prior to 'D' Day.

Sommerfield netting used on a number of advanced landing grounds was in such short supply tha it became normal practice to roll it up for use on the continent once our forces had gained a foothold.

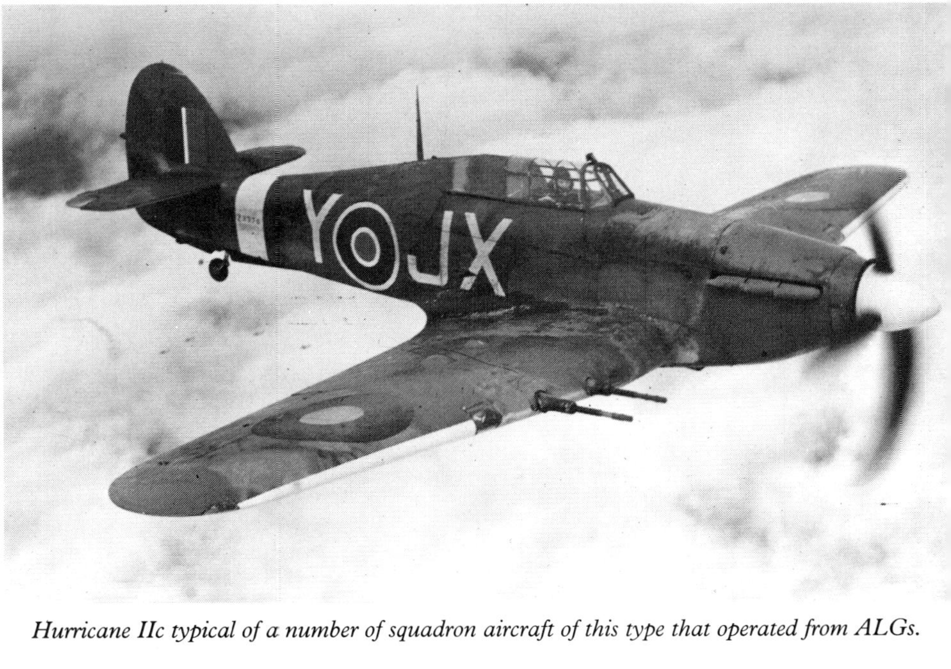

Hurricane IIc typical of a number of squadron aircraft of this type that operated from ALGs.

257 Squadron at Needs Ore Point 2 days prior to invasion of Normandy in 1944.

Fairey Abacore, a type that actually operated from a few of the ALGs to give smoke screen for our naval forces on 'D' Day.

LANDING GROUNDS USED

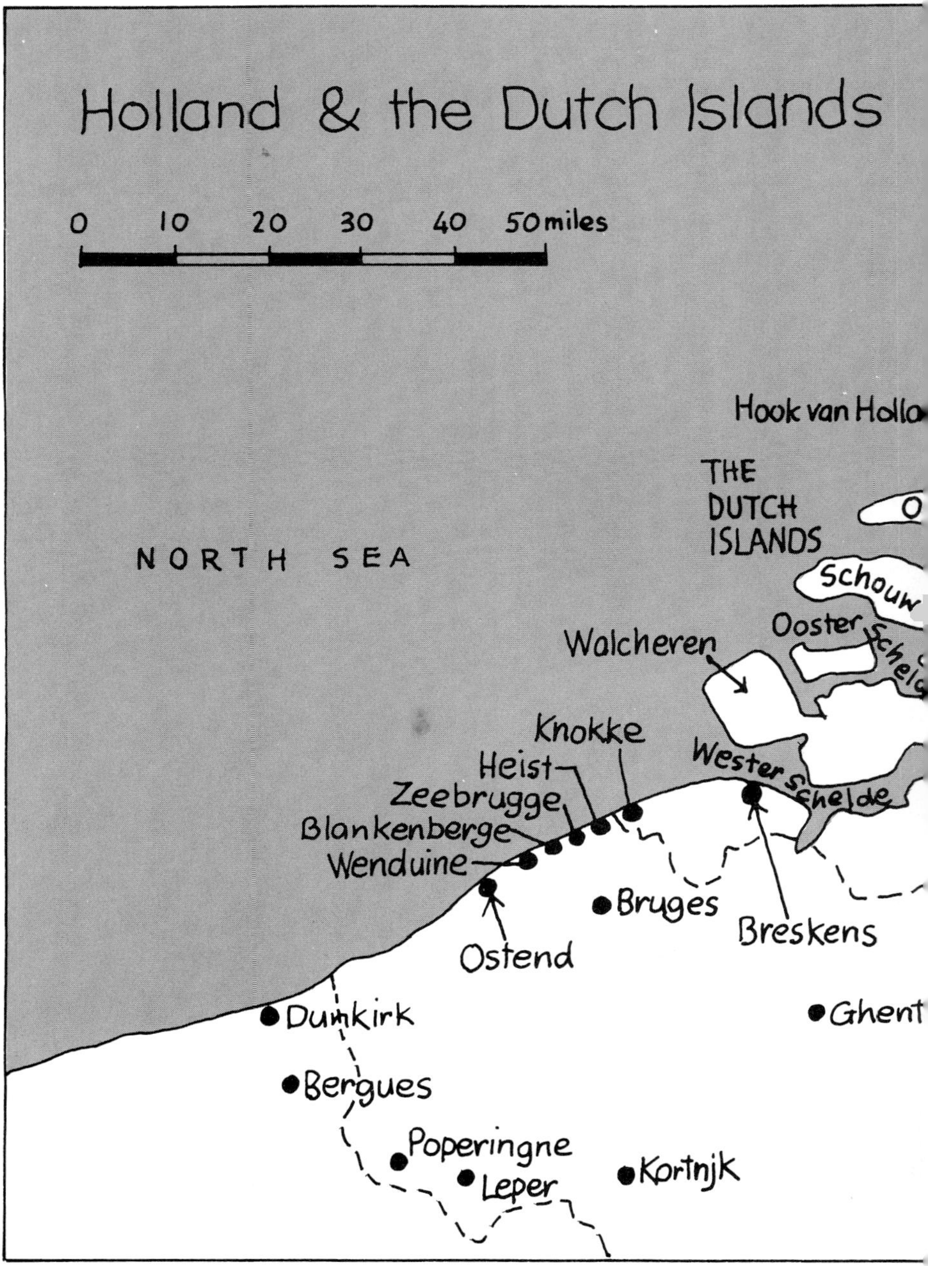

LLIED FORCES AFTER 'D' DAY

Christchurch (Somerton) Airfield from the air looking approx. North East. Housing developr and Airspeed factory indicates photograph to about early 1960s.

Photo credit: Portsmouth City Museum.

PICTURE INDEX